Air-Ice-Ocean Interaction

Miles McPhee

Air-Ice-Ocean Interaction

Turbulent Ocean Boundary Layer
Exchange Processes

 Springer

Miles McPhee
McPhee Research Company
Naches, WA
USA

ISBN 978-0-387-78334-5   e-ISBN 978-0-387-78335-2

Library of Congress Control Number: 2008928012

All Rights Reserved
© 2008 Springer Science + Business Media B.V.
No part of this work may be reproduced, stored in a retrieval system, or transmitted in any form or by any means, electronic, mechanical, photocopying, microfilming, recording or otherwise, without written permission from the Publisher, with the exception of any material supplied specifically for the purpose of being entered and executed on a computer system, for exclusive use by the purchaser of the work.

Printed on acid-free paper

9 8 7 6 5 4 3 2 1

springer.com

Preface

The outline of this monograph is based loosely on a series of lectures delivered to students enrolled in the air–sea–ice interaction course at the University Center in Svalbard. Both the lectures and the book are products of over three decades of trying (sometimes struggling) to distill from measurements at nearly two dozen ice-station projects of varying size and complexity, a somewhat cogent view of how the boundary layer that develops between sea ice and the underlying ocean behaves. While from many other perspectives, polar oceanographic research is always challenging, from the point of view expressed in this book, its overriding advantage is that sea ice provides a platform for making turbulence measurements that are extremely difficult elsewhere. In other words, it is a superb natural laboratory for studying planetary boundary layers throughout their whole extent. Still, it is a laboratory that demands much of its users, and for useful results, almost always requires hard work, plus a high degree of cooperation and perseverance among many scientists and support personnel. For me, this has been one of the most rewarding aspects of spending most of a scientific career devoted to high-latitude studies. Time and again, I have seen the harsh environment bring out the best of people in terms of inventiveness and all around competence. And the best is often very, very good. It would be a disservice to the many who have helped with the measurements, analysis, theory, and funding, to single out individuals for special thanks—it has always been a completely collaborative effort. The one exception is that I wish to thank my wife, Saundra, for her patience with the immediate task of my writing this book, for her occasional help with the field work, for maintaining a close-knit family during my frequent absences on remote projects, but mainly for always supporting and sharing my passion for the scientific endeavor.

Naches, Washington USA *Miles McPhee*
2008

Contents

1 **Introduction** 1
  1.1 Arctic Change 1
  1.2 The Southern Ocean 7
  1.3 Ekman's Seminal Paper 8
  1.4 Polar Boundary-Layer Field Studies 10
  1.5 Roadmap 12
  References 13

2 **Basic Physical Concepts** 15
  2.1 Conservation Equations in Fluids 15
  2.2 Reynolds Fluxes 16
  2.3 Rotation: The Coriolis Force and Geostrophy 18
    2.3.1 Geostrophic Shear 19
  2.4 Boundary-Layer Equations 19
  2.5 Inertial Oscillations 20
  2.6 Ekman Pumping 24
  2.7 The Equation of State for Seawater 28
  References 36

3 **Turbulence Basics** 39
  3.1 General Characteristics 39
  3.2 IOBL Measurement Techniques and Examples 40
    3.2.1 Smith Rotors 40
    3.2.2 Turbulence Instrument Clusters 41
    3.2.3 Momentum and Scalar Flux Measurements 43
    3.2.4 Estimating Confidence Limits for Covariance Calculations . 46
    3.2.5 Averaging Time and the Spectral Gap 48
  3.3 Turbulent Kinetic Energy Equation 51
  3.4 Scalar Variance Conservation 53
  3.5 Turbulence Spectra and the Energy Cascade 54
  3.6 Mixing Length, Eddy Viscosity, and the w Spectrum 58

vii

	3.7	Scalar Spectra	60
	References		62
4	**Similarity for the Ice/Ocean Boundary Layer**		**65**
	4.1	The Surface Layer	65
		4.1.1 Mixing Length in the Neutral Surface Layer	67
		4.1.2 The Law of the Wall and Surface Roughness Length	67
		4.1.3 Monin-Obukhov Similarity	68
	4.2	The Outer Layer	70
		4.2.1 Similarity for Turbulent Stress in the Outer Layer	72
		4.2.2 Rossby Similarity for the Neutral IOBL	74
		4.2.3 Similarity for the Stably Stratified IOBL	77
	4.3	IOBL Similarity and the Atmospheric Boundary Layer	81
		4.3.1 Dimensionless Shear	81
		4.3.2 The Rossby-Similarity Parameters for Stable Stratification	83
	4.4	Ice-Edge Bands	83
	References		85
5	**Turbulence Scales for the Ice/Ocean Boundary Layer**		**87**
	5.1	Neutral OBL Scales	87
		5.1.1 Ice Station Weddell	87
		5.1.2 Ice Station Polarstern	93
	5.2	The IOBL with Stabilizing Boundary Buoyancy Flux	95
	5.3	The Statically Unstable IOBL	98
	5.4	Velocity Scales in the IOBL	102
	5.5	Summary of IOBL Scales	105
	References		107
6	**The Ice/Ocean Interface**		**109**
	6.1	Enthalpy and Salt Balance at the Interface	110
	6.2	Turbulent Exchange Coefficients	112
	6.3	The "Three-Equation" Interface Solution	114
	6.4	Heat Flux Measurements and the Stanton Number for Sea Ice	116
	6.5	Double Diffusion—Melting	118
	6.6	Double Diffusion and False Bottoms	119
	6.7	Freezing—Is Double Diffusion Important?	125
	References		130
7	**A Numerical Model for the Ice/Ocean Boundary Layer**		**133**
	7.1	Difference Equations	133
	7.2	Boundary Conditions	135
		7.2.1 Flux of Variable θ Specified at Upper Surface	136
		7.2.2 Variable θ Specified at Upper Surface	136
		7.2.3 Dynamic Momentum Flux Condition	137
		7.2.4 Flux of θ Specified at the Bottom of Model Domain	138
		7.2.5 θ Specified at the Bottom of the Model Domain	138

	7.3	Steady-State Momentum Equation	139
	7.4	Distributed Sources	139
	7.5	Solution Technique	140
	7.6	The Local Turbulence Closure Model	140
	7.7	The Ice/Ocean Interface Submodel	143
		References	143
8		**LTC Modeling Examples**	145
	8.1	Diurnal Heating Near the Solstice, SHEBA	146
	8.2	Inertial Oscillations in Late Summer, SHEBA	151
		8.2.1 Wind Forced Model	151
		8.2.2 Models Forced by Surface Velocity	157
		8.2.3 Short-Term Velocity Prediction	160
	8.3	Marginal Static Stability, MaudNESS	162
		References	170
9		**The Steady Local Turbulence Closure Model**	173
	9.1	Model Description	176
	9.2	The Eddy Viscosity/Diffusivity Iteration	177
	9.3	Applications	184
		9.3.1 Ice Station Polarstern	184
		9.3.2 Underice Hydraulic Roughness for SHEBA	186
		9.3.3 SHEBA Time Series	189
		References	192
Colour Plate Section			193
Index			213

Chapter 1
Introduction

Abstract: Earth's polar regions play a pivotal role in climate, both as an important mediator in exchanges between the atmosphere and global ocean, and as a harbinger of climate change. Central to this impact is a thin layer of sea ice that is predominantly seasonal, and at maximum extent (at the end of austral winter) covers roughly 8% of the world ocean area. Sea ice affects climate in several important ways. It effectively insulates the ocean from the cold polar atmosphere, reducing both outgoing longwave radiation and convective heat exchange; it reflects a much higher proportion of incoming shortwave radiation than does open water; and by rejecting salt as it freezes, is capable of producing the cold, saline water that constitutes the end point in mixing processes that determine the density of the abyssal ocean. This chapter discusses the rationale behind a monograph on how sea ice affects atmosphere-ocean exchanges and how studies of turbulent exchange in the ice-ocean boundary layer have revealed much about how planetary boundary layers (where rotation is important) work in general. It briefly describes pertinent ice-station exercises, and lays out the framework for subsequent chapters.

1.1 Arctic Change

In early October 1997, I was part of a large party of polar scientists establishing the Surface Heat Budget of the Arctic (SHEBA) ice station, which was slated to drift for a year in multiyear pack ice over the Canadian Basin in the western Arctic Ocean. Its purpose was to assess the various energy components responsible for maintaining the perennial sea ice of the Arctic. Earlier, after a long week waiting in Tuktoyaktuk, NWT, for decent flying weather, most of us had boarded the Canadian Coast Guard icebreaker *Des Grosiellier*, which was to serve as our drifting base for the next year, anticipating an arduous trip following our escort icebreaker (the CCS *Louis St. Laurent*) as it battered its way through thick ice toward the center of the Beaufort Gyre. Instead, those of us who had been in that part of the Arctic before were astonished at how easy our passage was in ice often a meter or less thick. After a long search during which we began to wonder if we would even find a floe with a decent chance of surviving through the next summer, we eventually settled on ice about 2 m thick, and began the intense activity of deploying the various instrument systems.

Some time during that first week on the ice, I sat for an interview with reporters from a national news network, and during the questioning tried to succinctly voice the long practiced talking points summarizing the rationale for this complex and expensive experiment: that climate models were sensitive to the insulating and albedo properties of sea ice; that polar regions were the "canary in the mine" for global warming; that many of us suspected, even if we could not prove it unequivocally, that warming climate was tied to our collective appetite for fossil carbon, etc. The interview was long, and of course, I was anxious to get back to the task of putting my instruments in the water, yet I also understood the necessity of communicating our work and why we thought it important.[1]

Within a couple of days after the media interview, we were able to start the SHEBA ocean profiler and to obtain the first samples of upper ocean temperature and salinity. That evening I looked over the initial profiles, and compared them with data taken from a nearby location at the same time of year during the Arctic Ice Dynamics Joint Experiment (AIDJEX) in 1975. I had trouble sleeping that night, returning again and again to the thought that my children or at least their children would live to see an Arctic free of perennial sea ice. What kept me awake was a nebulous juxtaposition of the intellectual versus emotional sides of being a scientist. Just days before I had tried to explain to what might conceivably be a television audience of millions of people how our craving for fossil fuels could, according to the best models we had, permanently change at least one of the distinguishing polar caps of our planet, and that our mission here was to try to better understand what role rather esoteric processes like the albedo feedback might play in that transition. The analytical side of me understood and accepted those physical arguments, yet there was still something essentially conservative in me that rejected the idea that my species could really modify something as fundamental as the earth's climate in the short span of a few decades. But here was evidence that things were indeed changing, perhaps much faster than we had thought possible.

Why those initial profiles triggered my concern requires some explanation. First, they indicated that the upper ocean near the center of the Beaufort Gyre was at least 10% fresher than it had been in any previous measurements I had seen. Data from the central part of the Gyre are quite limited because compact and thick multiyear sea ice had traditionally made it one of the hardest Arctic regions to sample. By itself, the freshening was somewhat alarming but could have been due to a variety of factors, including an accumulation over several years of fresh water into the region from enhanced continental runoff combined with changes in the wind-driven circulation. However, from studying upper ocean evolution at the four AIDJEX drift

---

[1] When the interview finally aired later that fall on one of the network morning shows, it lasted for less than 1 min, and comprised mainly my communicating that, yes, I had accidentally fallen in the Arctic Ocean before, and that, yes, it was cold... My experience with other media interviews and conversations with journalists during the SHEBA deployment was in general much more positive.

1.1 Arctic Change

stations during the summer of 1975, we knew that in the normal course of the annual melt cycle, fresh water from the surface would form a relatively shallow seasonal pycnocline (density gradient). Just below this would be a layer of water preserving approximately the same temperature and salinity characteristics as the previous winter's mixed layer. By calculating the change in salinity of the water column above this remnant layer, we could estimate the amount of fresh water added over the melt season. It was clear from that first SHEBA sample in early October that a remnant mixed layer existed from about 30–45 m depth, and that the seasonal mixed layer in October (widespread basal freezing had not yet started) was so fresh that there must have been excessive melt during the summer of 1997.[2] In other words, the evidence indicated that a strong ice-albedo feedback had kicked in, and as I lay in my bunk aboard the *Des Grosiellier* that night what had been up until then a theoretical exercise now seemed very real.

The concept of albedo feedback is easily grasped—basically ice is highly reflective of incoming shortwave (solar) radiation while open seawater absorbs nearly all of it, so more open water melts ice which creates more open water, and so on— yet in addition to seasonal changes in albedo of the sea ice itself (Perovich et al. 2002), there are many subtleties in the problem, including storage of sensible heat in the upper ocean, the rate at which ice melts in contact with above freezing water, protection of thin ice by collection of meltwater at the ice undersurface, and mixing of sensible heat from below the ice-ocean boundary layer (hereafter abbreviated IOBL).

From an earlier analysis of ocean-to-ice heat flux during the AIDJEX experiment in the summer of 1975 (Maykut and McPhee 1995), we had gained an appreciation for the impact of relatively small changes in ice concentration on the ice mass balance. AIDJEX comprised an array of four drift stations, three of which made a triangle with sides roughly 100 km long, surrounding a central station. Over the course of the summer, we found that integrated ocean heat flux at the easternmost station (Blue Fox) was about 200 MJ m^{-2} compared with about 150 MJ m^{-2} at the other stations. The difference, equivalent to roughly 20 cm of ice melt, nearly all accumulated during a ten-day period beginning about August 10, nearly two months after the summer solstice. A mosaic of aerial photographs covering the entire AIDJEX array taken fortuitously on August 18, revealed that there was considerably more open water in the vicinity of Blue Fox compared with the other stations. Apparently this "open window" (we estimated about 25% open water) during a relatively brief period in late summer was enough to increase ice melt by about a third compared with the more compact (10% open water) regions to the west.

In the decade following the 1997–1998 SHEBA deployment, most of the scientific community and indeed the mass media have become aware of profound changes occurring in the Arctic. With regard to sea ice, the most striking symptom

---

[2] We later quantified these ideas (McPhee et al. 1998) and suggested that during the 1997 summer, freshening equivalent to as much as 2.5 m of ice melt had occurred in the seasonal mixed layer in the SHEBA vicinity. We cited isotope evidence (courtesy D. Kadko) that water in the intermediate remnant layer had been in contact with the surface earlier in the summer, and ruled out advection of fresh runoff as the primary source because of our distance from the continental margin.

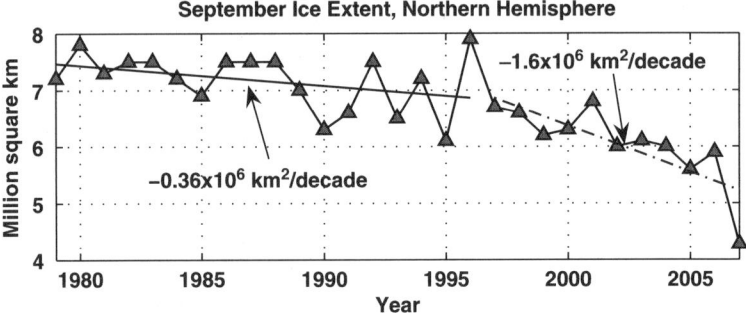

Fig. 1.1 Average Arctic ice extent in September (minimum) for the years 1979–2006, from the National Snow and Ice Data Center website. Two linear regressions are shown with slopes labeled for the period 1979–1996 and 1997–2007 (NSIDC data, Fetterer et al. 2007)

is the accelerating reduction in ice extent (Fig. 1.1), which since late 1978 has been accurately measured with satellite-borne passive microwave sensors. Note that despite the overall trend, during the earlier part of the satellite era, whenever a marked minimum occurred in September extent, it was nearly always followed in the next year by a significant rebound; e.g., the maximum September extent occurred in 1996 after the previous record minimum. Since 1997 this has not held, and a linear regression of minimum ice extent with time in the last decade shows a markedly increased rate of decline compared with the previous two decades. A continuation or acceleration of the rate of decrease obviously implies that the Arctic will be seasonally ice free by mid-century.

If ice-albedo feedback is a major factor in the retreat of the Arctic ice pack, then ice concentration during maximum sun elevation (June) should be a strong indicator of the state of the pack, perhaps a better diagnostic than minimum ice extent, since it is probably less subject to transient atmospheric conditions that can move ice near the marginal ice zones. The anomaly in total ice area relative to the average from 1979 to 2000 (Fig. 1.2) shows mostly the same trend as minimum extent (the correlation coefficient is 0.85), yet there is an interesting plateau in the eight years following 1997. While this suggests that perhaps the ice pack had reached some sort of equilibrium not obvious in the minimum ice extent records, in 2006 and again in 2007, the area has dipped precipitously, with nearly a million square kilometers more open water in June now than during the plateau period.[3] Possibly more ominous from the standpoint of survival of the summer ice pack are modeling estimates by Rigor and Wallace (2004) and more recently from satellite scatterometer data (Nghiem et al. 2006) that the area covered by perennial ice has decreased by as much as 2.5 million square kilometers since 1958. One cannot help but wonder what, if any, processes could reverse these important trends.

In some parts of the Arctic, changes in the ice cover appear to be accompanied by changes in the temperature and salinity structure of the upper ocean that are no less

---

[3] For perspective, the area of Alaska is about 1.7 million square kilometers.

1.1 Arctic Change 5

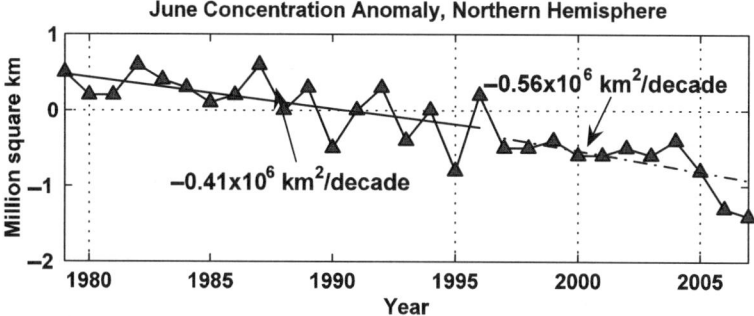

Fig. 1.2 Trend in total ice area in the Arctic during June relative to the average from 1979 to 2000 (10.3×10^6 km^2). Data from the NSIDC website, adjusted by assuming that concentration in the satellite "pole hole" centered around the north pole (different areas depending on the satellite) is complete (NSIDC data, Fetterer et al. 2007)

startling. In recent years, deployment of Ice Tethered Profilers (ITPs, Krischfield et al. in press, J. Atmos. Ocean. Technol.) have provided T/S profiles in the upper few hundred meters of the Arctic Ocean in near real time all year at several locations. In 2006 and 2007, buoys located in the Beaufort Gyre have shown remarkably fresh (and warm) water in the upper 100 m or so compared with conditions observed in 1975 during AIDJEX. The profiles labeled AJX in Fig. 1.3a and b were obtained by averaging all data gathered in September 1975, at AIDJEX station Snowbird, which was the northern most of the four manned stations. ITP6 was about 250 km north of the mean position of Snowbird.[4] Assuming that horizontal gradients were small in 1975, the difference in temperature (Fig. 1.3c) implies a change of heat content in the upper 100 m of as much as 330 MJ m^{-2}. This is enough heat to melt (or prevent formation of) well over a meter of sea ice if it were all somehow extracted at the surface. The decrease in salt content is, if anything, more significant. Again, using ice as a convenient measuring stick, the dilution implied by these profiles is equivalent to nearly 10 m of ice melt. This is clearly a major change in upper-ocean structure in this part of the world.

The Arctic Ocean resides over a mostly enclosed basin (it is often described a Mediterranean sea) bordered by some of the broadest continental shelves found anywhere. The dominant anticyclonic atmospheric circulation tends to collect fresh runoff and sea ice produced on the shelves inward toward the center at the surface; at the same time as a layer of cold, more saline water separating the surface waters from the underlying warmer water of Atlantic origin is maintained by drainage of cold brine resulting from freezing on the shelves (Aagaard and Coachman 1981). Thus the Arctic IOBL over much of the Arctic basin is bounded below by a very strong density gradient (pycnocline) that effectively limits interaction between the combined sea-ice/IOBL system and the underlying ocean. In the central Beaufort Gyre this barrier appears to be strengthening, presenting

---

[4] The September 1975, positions of the AIDJEX manned stations were either in open water or very near the ice edge in September 2007.

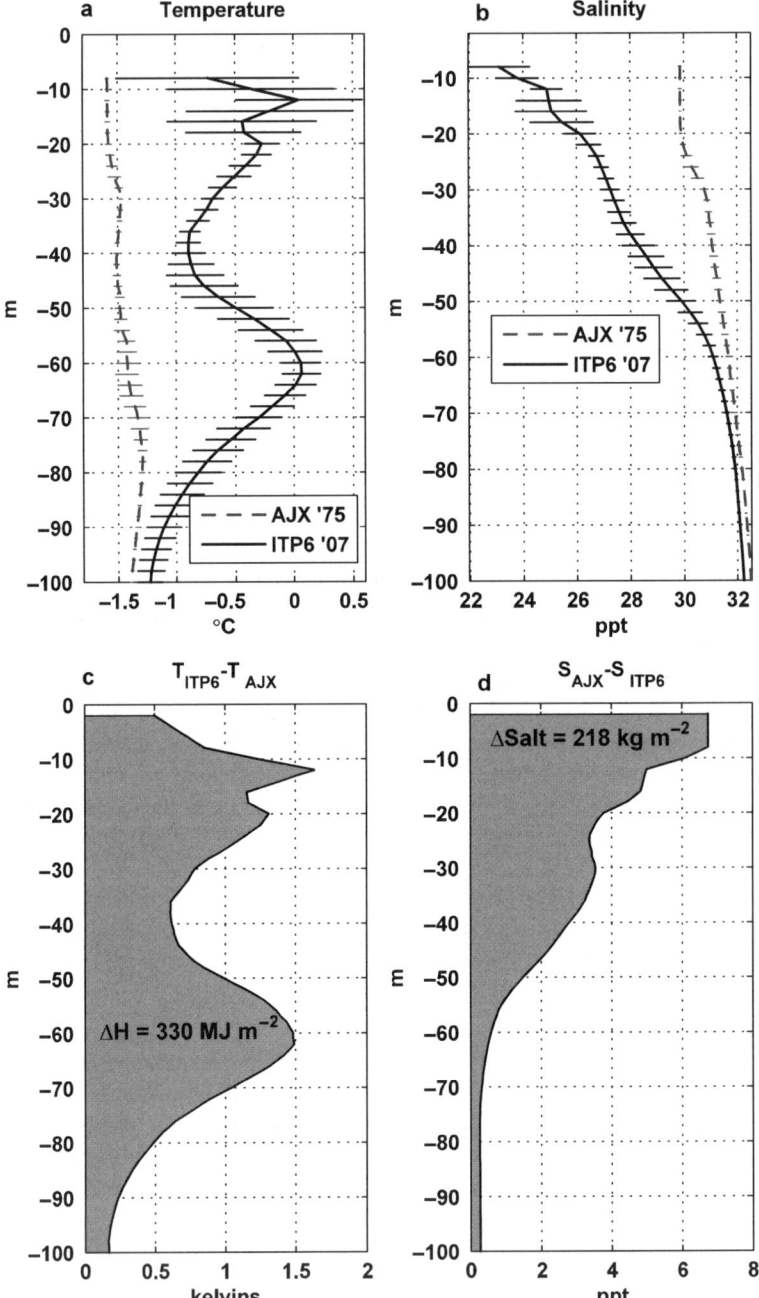

Fig. 1.3 Comparison of September average temperature **a** and salinity **b** in the upper 100 m of the water column at AIDJEX station Snowbird in 1975 (29 profiles, mean position: 74°13′ N, 141°22′ W) and from the Ice-Tethered Profiler buoy number 6 in 2007 (58 profiles, mean position: 76°37′ N, 142°58′ W. Horizontal bars indicate ± one standard deviation. **c** Change in temperature with total change in heat content. **d** Decrease in salinity, with equivalent change in salt content

an as yet unresolved competition between more heat available near the surface from summer insolation, but stronger stratification inhibiting the upward mixing of this heat.

1.2 The Southern Ocean

Sea ice extent in the Antarctic has not exhibited the marked decrease seen in the north; however, the Southern Ocean sea-ice pack is largely seasonal to begin with, and presents a whole set of interesting contrasts with the perennial pack of the Arctic. The Southern Ocean is completely unbounded by land to the north, and in general terms the westerly winds near the winter ice edge transport ice (and its fresh water) northward, leaving the Antarctic IOBL much saltier than its Arctic counterpart, with much less density contrast between it and the underlying warm deep water (WDW).

A consequence of weak stratification in the ocean around Antarctica, particularly in the Weddell Sea, is more direct communication between the surface and the deep ocean. In the eastern part of the Weddell Gyre (near the Prime Meridian), even in late winter the ice pack may be less than half a meter thick. Given atmospheric conditions influenced by the Antarctic continent, this requires on average an upward heat flux from the WDW of between 25 and 40 W m^{-2}, at least an order of magnitude greater than in the Arctic (Parkinson and Washington 1979; Schlosser et al. 1990; Gordon 1991; McPhee et al. 1999).

The lack of stratification also admits the possibility of deep reaching convection over the deep basin, with massive direct exchange of heat and other contaminants between the abyssal ocean and the atmosphere (one of the few places on earth where this is possible). In the mid-1970s, shortly after the first availability of satellite microwave imagery in the Southern Ocean, a large expanse of open water (or low concentration sea ice) was observed that remained for several winter seasons, well within the limits of the seasonal ice pack, and far from the continental shelves (Carsey 1980; Martinson et al. 1981). This persistent opening, called the Weddell Polynya, suggested deep convection that brought enough heat to the surface to prevent ice formation all winter over an area approaching 10% of the normal seasonal ice pack in the Weddell. Field observations confirmed the presence of a homogeneous "chimney" extending from the surface to 4,000 m (Gordon 1981) and extensive cooling of WDW during the Weddell Polynya years (Foldvik et al. 1985; McPhee 2003). Gordon (1991) interpreted the Polynya as manifestation of a "thermal mode" where sea ice formation is prevented by upward ocean heat flux and convection is driven by cooling alone. The presence of sea ice and its interaction with WDW through the IOBL, discussed in Section 8.3, are critical elements in determining what conditions might again initiate return to the thermal mode in the Weddell or elsewhere in the Southern Ocean. Were it to become widespread it would almost certainly have climate impact by venting massive amounts of heat, carbon dioxide and other contaminants from the deep ocean. Indeed, it is natural to speculate that the lockstep variations in proxy temperature and CO_2 from air bubbles trapped in the Vostok ice core spanning several glacial maxima and minima (Petit et al. 1999) result from interaction between the physics of deep convection in the Southern Ocean and other elements of the global carbon cycle (Sigman and Boyle 2000).

In both the Arctic and Antarctic, understanding how sea ice interfaces with the ocean through the medium of the IOBL is key to determining how polar regions interact with the climate. As implied above, the form that interaction takes depends strongly on the exchange of heat, salt and momentum at the immediate ice/ocean boundary; the density structure of the water column; and how those factors influence scales of turbulence in the IOBL.

1.3 Ekman's Seminal Paper

Any work on the IOBL footnotes to a remarkable paper published (the English version) in 1905 by V. W. Ekman, a Swedish oceanography student who worked with V. Bjerknes and F. Nansen on a mathematical theory to explain Nansen's observations that the *Fram*, on its famous 1893–1896 expedition in the Arctic Ocean, often drifted to the right of the surface wind. In the introduction to the paper, Ekman wrote

> On studying the observations of wind and ice-drift taken during the drift of the FRAM, Fridtjof Nansen found that the drift produced by a given wind did not, according to the general opinion, follow the wind's direction but deviated 20°–40° to the right. He explained this deviation as an obvious consequence of the earth's rotation; and he concluded further that the water-layer immediately below the surface must have a somewhat greater deviation than the latter and so on, since every water-layer is put in motion by the layer immediately above, sweeping over it like a wind...

He then proceeded to develop an even more elegant mathematical description of the process described in his prose, culminating in the famous spiral structure in rotating boundary layers now named after him. In modern notation, his solution to the steady state equations for a boundary layer in a rotating reference frame (Chapter 2) may be written in terms of a complex number representing a two-dimensional (horizontal) velocity vector:

$$V = V_x e_x + V_y e_y = V_r + i V_i :$$

$$V(z) = (i/fK)^{1/2} \tau_0 e^{(f/2K)^{1/2}(1+i)z} \tag{1.1}$$

where f is the Coriolis parameter (twice the local vertical rotation rate), K is "eddy viscosity"; τ_0 is kinematic stress at the boundary, and z is positive upward (negative in the ocean). Ekman noted that at the boundary $(z=0)$, the velocity and stress are related by a factor containing the term $(e^{i\pi/2})^{1/2} = (1/\sqrt{2})(1+i)$ which rotates surface velocity 45° to the right of surface stress in the northern hemisphere, qualitatively explaining Nansen's observations. The exponential term in (1.1) both

1.3 Ekman's Seminal Paper

attenuates velocity magnitude with increasing depth and rotates it clockwise (*cum sole*) relative to the surface. At the depth $d_E = \pi\sqrt{\frac{2K}{f}}$, the velocity is (somewhat counterintuitively) opposite the surface velocity. By careful reasoning, Ekman used this to infer values for K, and showed that it must be several orders of magnitude greater than molecular kinematic viscosity.

Although there was much indirect evidence for *cum sole* deflection of currents in both atmospheric and oceanic boundary layers, the first unequivocal example of an Ekman spiral was published by Hunkins (1966), who used a composite of current profiles measured over a two-month period at Arctic Drift Station Alpha during the International Geophysical Year in 1958 to fit an Ekman spiral starting a short distance below the ice/water interface, where the current (relative to geostrophic flow in the ocean) was 45° degrees from the interfacial stress. He inferred an eddy viscosity of about $0.0024\,\text{m}^2\,\text{s}^{-1}$ from the relatively small currents he measured. Ekman had suggested with remarkable insight (based partly on the setup of coastal currents during storms) that the eddy viscosity would depend on the square of the surface wind speed, i.e., on surface stress. In a footnote, he suggested that eddy viscosity would be roughly $0.0200\,\text{m}^2\,\text{s}^{-1}$ in a wind of $7\,\text{m}\,\text{s}^{-1}$, an order of magnitude greater than Hunkins's estimate, which was based on weak mean currents observed from ice. Nevertheless, the latter became a *de facto* standard for oceanographers for some time, apparently for lack of other definitive measurements. As described later, we now know that in essence Ekman got it right, and came very close to outlining the similarity concepts discussed in Chapter 4.

There is a great deal more in Ekman's (1905) paper than derivation of the steady-state Ekman spiral. With credit to Fredholm, he also presented a solution to the time-dependent problem and showed that circular currents oscillating with a period of a "half-pendulum day" (12 h at the poles) about the mean currents would be expected in the boundary layer. He reportedly sought somewhat unsuccessfully to measure inertial currents during his otherwise long and productive career. Now these circular currents are known to be ubiquitous in the ocean, and indeed often appear as cycloidal loops in ice drift trajectories (Section 2.5).

Ekman realized that ideally the integrated mass transport in the rotating boundary layer would be at right angle to the applied stress. This means, for example, that a southerly wind along a west coast in the northern hemisphere would drive surface water onshore, which would in time set up an onshore sea-surface tilt (coastal setup), driving a geostrophically adjusted current in the same direction as the wind. This current would in turn produce a bottom boundary layer, with offshore mass transport. A steady state could be achieved when the onshore wind driven transport at the surface balanced the offshore transport in the bottom boundary layer. With the opposite wind, bottom water would be driven onshore, balanced by offshore transport at the surface. This simple conceptual model explains much about how coastal sea level varies in response to wind, and why swimming off Oregon (or Capetown) beaches in summer, when there is a persistent northerly (southerly) wind, is not for the faint of heart.

Fortunately (for subsequent researchers) Ekman's (1905) paper lacked consideration of two important aspects of the IOBL. First, he did not consider that near the boundary, there would be more shear in the current than could be accounted for by vertically invariant eddy viscosity. Hunkins (1966) circumvented this problem by the reasonable expedient of assuming that stress in this shallow, enhanced shear layer would be nearly constant, and that the true Ekman layer began where the velocity and stress were 45° apart in direction, not right at the boundary. Ekman's second omission was neglect of how the scalar variables temperature and salinity that determine density were transported in the boundary layer, with sometimes very important impact on turbulence scales, hence eddy viscosity. Overall, though, writing a paper that described the basic physics of turbulent transfer in rotating boundary layers, foretold the presence of inertial oscillations some decades before instrumentation was available to measure them adequately, and laid the foundation for understanding coastal upwelling and storm surges was a remarkable feat, and in many respects far ahead of its time. At least from the perspective of its influence on geophysical fluid dynamics, the paper perhaps ranks with other famous scientific papers published in 1905, and it all began from the most obvious manifestation of the IOBL: sea ice does not drift directly down wind.

1.4 Polar Boundary-Layer Field Studies

My introduction to polar studies was mostly serendipitous, stemming from a chance conversation about turbulence with a favorite professor (J. Dungan Smith) when I was a first-year graduate student in the geophysics program at the University of Washington in 1971. In March 1972, I thus found myself standing on the flight deck of a C130 Hercules as it made the first nighttime landing on a frozen-lead runway lit with smudge pots about 300 nm north of Barrow, Alaska. My memory of the remainder of that night is the roar (and smell) of C130 turbines as flight after flight landed, and all hands turned out to offload tons of scientific equipment and support materiel. The next day I had a chance to observe what was for me a completely new environment: breathtakingly cold in spite of dazzling sunlight; a terrain of pressure ridges and sastrugi, like miniature landforms; and a color spectrum consisting only of gradations from blue to white. In a fundamental way, I have enjoyed the polar environment ever since.

By any standard, the 1972 AIDJEX Pilot Study was an enormous undertaking. Organized under the leadership of N. Untersteiner, peak occupancy of the station exceeded 80 scientists and support personnel. It was supplied by eighteen C130 and numerous smaller aircraft flights (Heiberg and Bjornert 1972). The AIDJEX Pilot Study also provided me, through my association with J. D. Smith, an opportunity afforded few graduate students: access to an unprecedented data set with simultaneous measurements of turbulent stress and mean velocity at several levels through an entire rotational planetary boundary layer. What Smith realized, and what I came soon to appreciate, was that drifting ice stations provided superb laboratories for studying

1.4 Polar Boundary-Layer Field Studies

ocean boundary layer (OBL) physics. Without vertical platform (ship) motion and the complicating factor of surface gravity waves, it was relatively easy to measure small velocity fluctuations across much of the turbulent spectrum. Wind-driven ice typically drifts with the maximum velocity in OBL, and given its considerable momentum, it provides a remarkably steady platform from which to measure the small fluctuations in velocity and scalar contaminants that constitute turbulent exchange.

In the intervening years I have participated in more than 20 polar field programs in both hemispheres. In addition to descriptions of SHEBA and AIDJEX already provided, some of the projects that provided additional data used in this volume were:

Marginal Ice Zone Experiment (MIZEX, June–July 1984): A multinational, multi-ship project in the Greenland Sea/Fram Strait region. We made measurements from two different floes drifting near the ice edge, supported by the M/V *Polar Queen*. Late in the project, northerly winds blew our floe across an upper-ocean temperature front marking the boundary of an eddy identified later in satellite imagery. After crossing into the warmer water, turbulent heat flux increased dramatically (Section 5.2).

Coordinated Eastern Arctic Experiment (CEAREX, March–April 1989): A late winter project north of Fram Strait established by aircraft. It mostly drifted along the NW flank of Yermak Plateau, and was notable in that in contrast to most of the Arctic Ocean experiments, the main driving was tidal and internal ice stress gradients rather than wind, the mixed layer was relatively deep, with no underlying cold, saline layer.

Ice Station Weddell (ISW, February–April 1992): A drift station following closely the track of HMS *Endurance* in 1915–1916 east of the Antarctic Peninsula (Gordon et al. 1993). The station was deployed by the Soviet research ice breaker R/V *Federov*, and recovered during the maiden voyage of the R/V *Nathaniel B. Palmer*, chartered for the US National Science Foundation. My turbulence apparatus was deployed and operated by R. Andersen and D. Martinson.

The **Lead Experiment** (LeadEX, March–April 1992): An ambitious experiment designed to move an entire ice camp by helicopters and/or snow machines to the edges of freezing leads within hours of the lead opening. The main station was deployed by air about 300 km NNE of Prudhoe Bay, Alaska. We deployed to four different leads (Section 5.3 and Fig. 5.12).

The **Antarctic Zone Flux Experiment** (ANZFLUX, June–August 1994): Winter experiment in the eastern Weddell Sea, with two drifts, one west of and the other over Maud Rise (a seamount centered near 65° S, 3° E.) In several storms we encountered extreme conditions of stress and heat flux in the upper ocean (McPhee et al. 1996).

Ice Station Polarstern (ISPOL, November 2004–January 2005): A drift experiment in early summer near the track of ISW (and the *Endurance*), supported by the German research icebreaker, R/V *Polarstern*. We drifted with a relatively large heterogeneous floe made up of multiyear and first-year ice fragments. in the western Weddell (Hellmer et al. 2006), forced by a combination of wind, tides, and mean flow.

Maud Rise Nonlinear Equation of State Study (MaudNESS, July–September 2005): A winter experiment in the eastern Weddell Sea designed to study upper ocean mixing in a low stability environment, from the R/V *Nathaniel B. Palmer*. The experiment comprised a rapid survey of upper ocean properties above and on the flanks of the Maud Rise seamount, two ship-supported drifts with various instrumentation on the ship and adjacent floe, and a series of short drifts with all instrumentation, including turbulence measuring equipment deployed from the ship.

Svalbard Fjord Studies An ongoing series of short field studies from fast ice in Svalbard fjords, usually done in collaboration with the University Center in Svalbard (UNIS). By measuring turbulence during tidal cycles along with ice characteristics, these experiments were designed to look at specific aspects of heat and salt exchange near the ice/ocean interface.

1.5 Roadmap

This book is intended to serve two purposes. First, it strives to summarize our present understanding of how sea ice and the upper ocean interact, and how that understanding may be applied in models that predict future changes. Rapid and apparently accelerating changes in the state of the Arctic ice pack lend a sense of urgency: essentially, it seems that Nature is solving the equations a lot faster than we are.

The second major aim is to consolidate what the rather unique measurements made from the ice-platform "laboratory" imply about the scales of turbulence and how fluid boundary layers work when rotation and buoyancy flux are important.

Chapter 2 is a quick review of basic fluid dynamical principles, with emphasis on the upper ocean and planetary boundary layer. Chapter 3 is a somewhat cursory summary of turbulence principles, with emphasis on direct measurements of covariance statistics as estimators of turbulent fluxes in the IOBL, and the use of turbulence spectra to infer features of the turbulent kinetic energy cascade and dominant scales. Chapter 4 explores concepts of fluid dynamical similarity as they apply to the IOBL, and emphasizes connections between the oceanic and atmospheric boundary layers. Chapter 5 is the "observational meat" of the work, using measurements to assess the impact of stress and rotation, plus stabilizing and destabilizing buoyancy flux, on scales of turbulence in the IOBL. Chapter 6 explores the small-scale processes that govern the transfer of heat, salt, and mass across the immediate boundary at the ice/ocean interface, including double-diffusive effects when ice is melting. Chapter 7 introduces a fairly standard, one (spatial) dimensional numerical model solution approach, along with an algorithm for implementing a first-order local-turbulence-closure (LTC) technique incorporating the similarity and scaling arguments of Chapters 5 and 6. Chapter 8 exercises the time-dependent numerical in three examples, chosen to illustrate (i) absorption and distribution of solar energy in the IOBL near summer solstice; (ii) inertial response of the IOBL to rapid changes in surface forcing; and (iii) mixing in an IOBL bounded below by a combination of strong thermocline and weak halocline, resulting in very little density contrast between the upper cold layer, and the underlying warm layer. Finally, Chapter 9 explores using a steady-state version of the IOBL model to "scale up" measurements made in a particular location on a floe (often chosen for ease of deployment) to be representative of the entire floe or surrounding region.

References

Aagaard, K., Coachman, L. K., and Carmack, E. C.: On the halocline of the Arctic Ocean. Deep-Sea Res., 28, 529–545 (1981)
Carsey, F.: Microwave observations of the Weddell Polynya. Mon. Wea. Rev., 108, 2032–2044 (1980)
Ekman, V. W.: On the influence of the earth's rotation on ocean currents. Ark. Mat. Astr. Fys., 2, 1–52 (1905)
Fetterer, F., Knowles, K., Meier, W., and Savoie, M.: Sea Ice Index, Boulder, CO, National Snow and Ice Data Center, Digital Media (2007)
Foldvik, A., Gammelsrød, T., and Tørresen, T.: Hydrographic observations from the Weddell Sea during the Norwegian Antarctic Research Expedition 1976/77. Polar Research 3 n.s., 177–193 (1985)
Gordon, A. L.: Seasonality of Southern Ocean sea ice. J. Geophys. Res., 85, 4193–4197 (1981)
Gordon, A. L.: Two stable modes of Southern Ocean winter stratification. In: Deep Convection and Deep Water Formation in the Oceans, Chu, P.-C. and Gascard, J.-C. (eds.) Elsevier Oceanography Series, 57, pp. 17–35. Elsevier, Amsterdam (1991)
Gordon, A. L. and Ice Station Weddell Group of Principal Investigators and Chief Scientists: Weddell Sea Exploration from Ice Station. EOS, Trans. Am. Geophys. Union, 74, 121–126 (1993)
Heiberg, A. and Bjornert, R.: Operations and logistics support, 1972 AIDJEX Pilot Study. AIDJEX Bull., 14, 1–11 (1972)
Hellmer, H. H., Haas, C., Dieckmann, G. S., and Schröder, M.: Sea Ice Feedbacks Observed in Western Weddell Sea. EOS, Trans. Am. Geophys. Union, 87 (18), 173, 179 (2006)
Hunkins, K.: Ekman drift currents in the Arctic Ocean. Deep-Sea Res., 13, 607–620 (1966)
Martinson, D. G., Killworth, P. D., and Gordon, A. L.: A convective model for the Weddell Polynya. J. Phys. Oceanogr., 11, 466–488 (1981)
Maykut, G. A. and McPhee, M. G.: Solar heating of the Arctic mixed layer. J. Geophys. Res., 100, 24691–24703 (1995)
McPhee, M. G.: Is thermobaricity a major factor in Southern Ocean ventilation? Antarctic Science, 15(1), 153–160 (2003)
McPhee M. G., Ackley, S. F., Guest, P., Huber, B. A., Martinson, D. G., Morison, J. H., Muench, R. D., Padman, L., and Stanton, T. P.: The Antarctic Zone Flux Experiment. Bull. Am. Met. Soc., 77, 1221–1232 (1996)
McPhee, M. G., Stanton, T. P., Morison, J. H., and Martinson, D.: Freshening of the upper ocean in the Arctic: Is perennial sea ice disappearing?. Geophys. Res. Lett., 25, 1729–1732 (1998)
McPhee, M. G., Kottmeier, C., and Morison, J. H.: Ocean heat flux in the central Weddell Sea in winter. J. Phys. Oceanogr., 29, 1166–1179 (1999)
Nghiem, S. V., Chao, Y., Neumann, G., Li, P., Perovich, D. K., Street, T., and Clemente-Colon, P.: Depletion of perennial sea ice in the East Arctic Region. Geophys. Res. Lett., 33, L17501 (2006), doi: 10.1029/2006GL027198
Parkinson, C. L. and Washington, W. M.: A large-scale numerical model of sea ice. J. Geophys. Res., 84, 311–337 (1979)

Perovich, D. K., Tucker, W. B., III, and Ligett, K. A.: Aerial observations of the evolution of ice surface conditions during summer. J. Geophys. Res., 107 (C10), 8048 (2002), doi: 10.1029/2000JC000449

Petit, J. R., et al.: Climate and atmospheric history of the past 420,000 years from the Vostok ice core. Nature, 399, 429–436 (1999)

Rigor, I. G. and Wallace, J. M.: Variations in the age of Arctic sea-ice and summer sea-ice extent. Geophys. Res. Lett., 31, L09401 (2004), doi: 10.1029/2004GL019492

Schlosser, P., Bayer, R., Foldvik, A., Gammelsrød, T., Rohardt, G., and Munnich, K. O.: Oxygen 18 and helium as tracers of ice shelf water and water/ice interaction in the Weddell Sea. J. Geophys. Res., 95, 3253–3264 (1990)

Sigman, D. M. and Boyle, E. A.: Glacial/interglacial variations in atmospheric carbon dioxide. Nature, 407, 859–869 (2000)

Chapter 2
Basic Physical Concepts

Abstract: At high latitudes, two features of geophysical fluid dynamics are particularly apparent: first, the impact of rotation is stronger near the poles than elsewhere; and second, the combination of cold temperature and salt injection inherent in the freezing process produces very dense water, so that the polar and subpolar regions provide the main conduit by which the abyssal ocean communicates with the remainder of the climate system. A cursory review of some basic physical properties of ocean dynamics particularly relevant to the IOBL is presented in this chapter. More rigorous treatment may be found in standard geophysical fluid dynamics textbooks (e.g., Gill 1982; Pedlosky 1987).

2.1 Conservation Equations in Fluids

For an arbitrary property of a fluid, denoted here by γ, the net flux, F^γ, of γ across a closed surface enclosing a volume in the fluid is equal to the integral of the flux divergence over the volume, by Gauss's theorem

$$\iint_{\text{closed surface}} (F^\gamma \cdot \boldsymbol{n}) da = \iiint_{\text{volume}} \nabla \cdot F^\gamma dV \qquad (2.1)$$

The local time rate of change of γ is minus the net flux across a closed surface out of the body plus the integral of any internal sources/sinks, and from (2.1) the standard conservation equation is

$$\frac{\partial \gamma}{\partial t} = -\nabla \cdot F^\gamma + Q^\gamma \qquad (2.2)$$

For scalars the flux term is a vector and is often considered in two parts: a flux associated with molecular diffusivity, and an advective term:

$$F^\gamma = -k_\gamma \nabla \gamma + \gamma \boldsymbol{u} \qquad (2.3)$$

For geophysical boundary layer flows, the advective flux (i.e., how the quantity γ is carried with mean and turbulent fluctuations in the flow) almost always dominates, so $\boldsymbol{F}^\gamma \cong \gamma \boldsymbol{u}$. Using the incompressibility conditions, $\nabla \cdot \boldsymbol{u} = \boldsymbol{0}$, and ignoring molecular diffusion, (2.2) becomes

$$\frac{\partial \gamma}{\partial t} + \boldsymbol{u} \cdot \nabla \gamma = Q^\gamma \tag{2.4}$$

for an arbitrary scalar quantity. Scalar equations for heat and salt are of vital importance in the IOBL, but the principles hold for other contaminants as well.

Conservation of enthalpy may be expressed as an equation for temperature in the fluid:

$$\frac{\partial T}{\partial t} + \boldsymbol{u} \cdot \nabla T = Q^H / (\rho c_p) \tag{2.5}$$

where ρ is density, c_p is specific heat at constant pressure (close to 4×10^3 J kg^{-1}K^{-1} for seawater near freezing), and Q^H is the source term, which typically comprises solar radiation flux divergence in the upper part of the water column, but might also include, for example, a phase change associated with nucleation of frazil crystals in the water away from the immediate interface. The corresponding salt equation is

$$\frac{\partial S}{\partial t} + \boldsymbol{u} \cdot \nabla S = Q^S / \rho \tag{2.6}$$

where S is salinity expressed in units of the practical salinity scale (henceforth designated *psu*, corresponding closely to parts per thousand). As above a possible source Q^S within the fluid might arise from nucleation of frazil crystals.

Substitute vector momentum for the arbitrary property γ in (2.4), and interpret the "momentum source term" as the sum of a pressure gradient in the fluid, and the acceleration of gravity acting on small density perturbations in the fluid (the Boussinesq approximation), to arrive at Euler's equation (essentially Newton's 2nd law for fluids, ignoring molecular diffusion):

$$\frac{\partial \boldsymbol{u}}{\partial t} + \boldsymbol{u} \cdot \nabla \boldsymbol{u} = -\nabla p / \rho - g \frac{\rho'}{\rho} \boldsymbol{k} \tag{2.7}$$

The fact that molecular diffusion was ignored in arriving at (2.7) does not exclude the impact of friction in the fluid, because the nonlinear advective term $\boldsymbol{u} \cdot \nabla \boldsymbol{u}$ provides a link via turbulence between the large-scale flow and dissipative processes at small scales.

2.2 Reynolds Fluxes

A local velocity vector may be expressed as $\boldsymbol{u} = \langle \boldsymbol{u} \rangle + \boldsymbol{u}'$ where the angle bracket denotes an instantaneous ensemble average over some area large compared with the scale of the "energy containing" turbulent eddies in a flow; $\boldsymbol{u}'$ is the *deviatory*

2.2 Reynolds Fluxes

velocity, with zero mean. The average of products, say two different components of the instantaneous local flow, is

$$\langle u_i u_j \rangle = \langle u_i \rangle \langle u_j \rangle + \langle u_i' u_j' \rangle \tag{2.8}$$

and, for example, the average of the jth component of the advection term in the Euler equation (2.7) is[1]

$$\left\langle u_i \frac{\partial u_j}{\partial x_i} \right\rangle = \langle u_i \rangle \frac{\partial \langle u_j \rangle}{\partial x_i} + \frac{\partial \langle u_i' u_j' \rangle}{\partial x_i} \tag{2.9}$$

The last term is the contribution of turbulence to the momentum balance, and may be interpreted as a symmetric kinematic stress tensor (with units of velocity squared):

$$\tau_{ij} = -\langle u_i' u_j' \rangle \tag{2.10}$$

called the *Reynolds stress tensor*. Rigorously, the Reynolds stress decomposition implies steady flow, but practically it provides a useful approximation provided the time scale over which the mean flow varies is long compared with the scales of typical turbulent eddies effecting transfer in the flow, i.e., that a *spectral gap* between the mean flow and turbulence exists. Similarly, measurements are most often made as time series at particular locations in a flow as the fluid streams past, rather than as a "snapshot" over the entire domain at a particular moment. In this case, the frozen-field (Taylor's) hypothesis holds that the turbulence structure moving past the measurement site is representative of the spatial structure, and that the covariance (in time) matrix of deviatory velocity components is representative of the true Reynolds stress tensor. Note that Taylor's hypothesis requires that turbulent velocities be considerably smaller than the mean velocity (Tennekes and Lumley 1972), a condition sometimes violated in oceanic flows.

Dropping the angle brackets, let quantities without primes denote the "mean" values, allowed to vary with time under the stipulations noted above. The Boussinesq form of the Euler equation then become

$$\frac{\partial \boldsymbol{u}}{\partial t} + \boldsymbol{u} \cdot \nabla \boldsymbol{u} = -\nabla p / \rho - g \frac{\rho'}{\rho} \boldsymbol{k} + \nabla \cdot \underset{\sim}{\boldsymbol{\tau}} \tag{2.11}$$

where $\underset{\sim}{\tau}$ is the Reynolds stress tensor. Note that despite ignoring molecular viscosity in deriving (2.11), we have recovered friction in the form of a deviatory stress associated with nonlinear turbulent fluctuations from instabilities in the flow.

The terminology of Reynolds stress is often ambiguous—in most cases, we are concerned with the vertical variation of the horizontal deviatoric stress components, i.e.

---

[1] Where appropriate the Einstein summation convention is adopted, wherein repeated indices imply summation.

$$\frac{\partial \tau_{31}}{\partial x_3}e_1 + \frac{\partial \tau_{32}}{\partial x_3}e_2 = \frac{\partial \tau}{\partial z} = -\frac{\partial(\langle u'w'\rangle + i\langle v'w'\rangle)}{\partial z}$$

where τ is a horizontal traction vector, expressed here as a complex number $\tau = -\langle u'w'\rangle - i\langle v'w'\rangle$. This horizontal traction vector is often referred to as the *Reynolds stress*, but with the understanding that the complete description is a tensor. Note that the trace of the Reynolds stress tensor (sum of the diagonal elements) is twice the turbulent kinetic energy per unit mass.

2.3 Rotation: The Coriolis Force and Geostrophy

Let $r = (i, j, k)$ be a unit vector in a reference frame rotation with angular velocity ω. In an unaccelerated (inertial) frame, the time derivatives of the unit vectors are

$$\frac{di}{dt} = \omega \times i \qquad \frac{dj}{dt} = \omega \times j \qquad \frac{dk}{dt} = \omega \times k$$

Let $R = xi + yj + zk$ be a position vector in the rotating frame. Then differentiating R with time

$$\dot{R} = \dot{x}i + \dot{y}j + \dot{z}k + x(\omega \times i) + y(\omega \times j) + z(\omega \times k) = V + \omega \times R \qquad (2.12)$$

where V is the velocity in the rotating frame. One further differentiation in time expresses the acceleration in the rotating frame:

$$\ddot{R} = \dot{V} + 2\omega \times V + \omega \times (\omega \times R) \qquad (2.13)$$

The last term on the right is centripetal acceleration, usually incorporated into gravity if considered at all (it is small at high latitudes), whereas the second term on the RHS of (2.13) is the Coriolis acceleration, and is of paramount importance for many geophysical flows. The vertical component of the rotation vector acting on the horizontal component of flow is what we are commonly interested in, described by the Coriolis parameter, $f = 2\omega_3 = 2|\omega|\sin\phi$ where ϕ is latitude, with the convention that latitude is positive in the northern hemisphere and negative in the south. The inertial period is $2\pi/f$. At high latitudes, the Coriolis acceleration is strong (because ω_3 is large) and the effects of rotation are more pronounced than at lower latitudes. The angular rotation speed of the earth is $7.292 \times 10^{-5}\,\mathrm{s}^{-1}$, so the inertial period, $2\pi/f$, is 12.23 h at 78°N.

If the advective terms (including associated Reynolds stress) are ignored, the horizontal part of the Euler equation in a rotating reference frame becomes

$$\frac{\partial u}{\partial t} + f k \times u = -\nabla_H p/\rho \qquad (2.14)$$

Define *geostrophic current*, u_g, as the steady current that balances the horizontal pressure gradient. If rapid variation in air pressure is neglected, it may be expressed in terms of the gradient of sea-surface elevation:

$$f k \times u_g \equiv -g\nabla \eta \qquad (2.15)$$

2.3.1 Geostrophic Shear

Differentiating steady, geostrophic velocity (2.15) with respect to z, provides an equation for vertical shear in terms of horizontal density gradients:

$$f\mathbf{k} \times \frac{\partial \mathbf{u}_g}{\partial z} = -\frac{\nabla_H}{\rho}\left(\frac{\partial p}{\partial z}\right) + \frac{\nabla_H p}{\rho^2}\frac{\partial \rho}{\partial z} \qquad (2.16)$$

In nearly all practical applications the change in density with depth is small compared to its mean value, and the second term on the right is negligible. Thus with the hydrostatic equation, $\partial p/\partial z = -g\rho$, an expression for shear in geostrophic current is given to good approximation by

$$\frac{\partial \mathbf{u}_g}{\partial z} \doteq -\frac{g}{\rho f}\mathbf{k} \times \nabla_H \rho \qquad (2.17)$$

By this relation, vertical current shear may be present even in the absence of friction, potentially a significant factor as ice drifts across water with large horizontal salinity gradients. In the atmosphere, where density is controlled mainly by temperature, (2.17) defines the *thermal wind* relation, i.e., vertical wind shear and differential advection associated with temperature fronts.

2.4 Boundary-Layer Equations

With rotation, the turbulent Boussinesq equation (2.11) becomes

$$\frac{\partial \mathbf{u}}{\partial t} + \mathbf{u} \cdot \nabla \mathbf{u} + f\mathbf{k} \times \mathbf{u} = -\nabla p/\rho - g\frac{\rho'}{\rho}\mathbf{k} + \nabla \cdot \underset{\sim}{\tau} \qquad (2.18)$$

where the impact of the Coriolis force on the deviatory velocities is considered negligible in most IOBL applications, especially at high latitudes. In general, the horizontal components of (2.18) are of most interest. In the absence of wind stress or forcing from internal stress gradients, ice will respond to a horizontal pressure gradient from tilt of the sea; i.e., shear between it and the underlying ocean will vanish so that in a steady state the ice velocity is just $\mathbf{u}_g$. It is thus often convenient to consider boundary layer flow in a frame of reference translating with the mean geostrophic current, so that the pressure gradient can be eliminated from (2.18), yielding an equation for the horizontal components of relative velocity

$$\frac{\partial \mathbf{u}_r}{\partial t} + \mathbf{u}_r \cdot \nabla \mathbf{u}_r + f\mathbf{k} \times \mathbf{u}_r = \nabla \cdot \underset{\sim}{\tau} \qquad (2.19)$$

where $\mathbf{u}_r = \mathbf{u} - \mathbf{u}_g$. Unless otherwise noted, in what follows we will tacitly assume that IOBL velocity is velocity with respect to the undisturbed ocean flow and drop from the r subscript from (2.19).

An important implication of (2.19) is that the steady state volume transport (i.e., the depth integral of velocity) in the boundary layer relative to the geostrophic flow is at right angle (*cum sole*) to the surface stress and proportional to its magnitude. If horizontal velocity is expressed as a complex number $\boldsymbol{u} = u_x + iu_y$ the steady IOBL momentum equation is

$$if\boldsymbol{u} = \frac{\partial \tau}{\partial z} \qquad (2.20)$$

At some level near the far extent of the boundary layer, the turbulent stress is zero, so integrating (2.20) from that level to the surface provides

$$if \int_{z_{bl}}^{0} \boldsymbol{u}\,dz = if\boldsymbol{M} = \tau_0 \qquad (2.21)$$

where $\boldsymbol{M}$ is the vector volume transport and τ_0 is the kinematic stress at the boundary. Multiplying a horizontal vector by i rotates it by 90°, thus volume transport will be approximately perpendicular to surface stress, regardless of details of turbulence in the IOBL. However, a shallow layer will require higher mean velocity than a deep layer to effect the same transport, which places an important constraint on boundary layer scales.

2.5 Inertial Oscillations

Ekman (1905) in his classic paper, pointed out (with credit to Fredholm) the possibility of oscillations in the upper ocean having the inertial period $2\pi/f$. Heuristically, *inertial oscillations* are easily demonstrated by considering the time-dependent volume transport equation obtained by vertically integrating the horizontally homogeneous version of (2.19):

$$\frac{\partial \boldsymbol{M}}{\partial t} + if\boldsymbol{M} = \tau_0 \qquad (2.22)$$

Suppose an upper ocean system initially at rest is subjected to an impulsive stress in the y-direction, $i\tau_0$ at time $t = 0$. It is easily verified that the complex solution of (2.22) is

$$\boldsymbol{M} = \frac{\tau_0}{f}(1 - e^{-ift})$$

The solution, sketched in Fig. 2.1, traces a circle in one inertial period about the steady-state balance $\boldsymbol{M}_{ss} = \tau_0/f$, but because there is no friction in this system it continues to oscillate with the inertial period, averaging $\boldsymbol{M}_{ss}$, but never having the steady-state value. Despite the seeming unreality of this example, it is instructive to consider some numbers. A typical kinematic surface stress during a moderate squall might be $\tau_0 = 2 \times 10^{-4}\,\text{m}^2\,\text{s}^{-2}$, with a maximum volume transport (occurring at $t = 6, 18, 30\,\text{h}$, etc., at the North Pole) of about $2.75\,\text{m}^2\,\text{s}^{-1}$. If the summertime mixed layer was 25 m thick, the depth averaged velocity in the boundary layer would be around $11\,\text{cm}\,\text{s}^{-1}$.

2.5 Inertial Oscillations

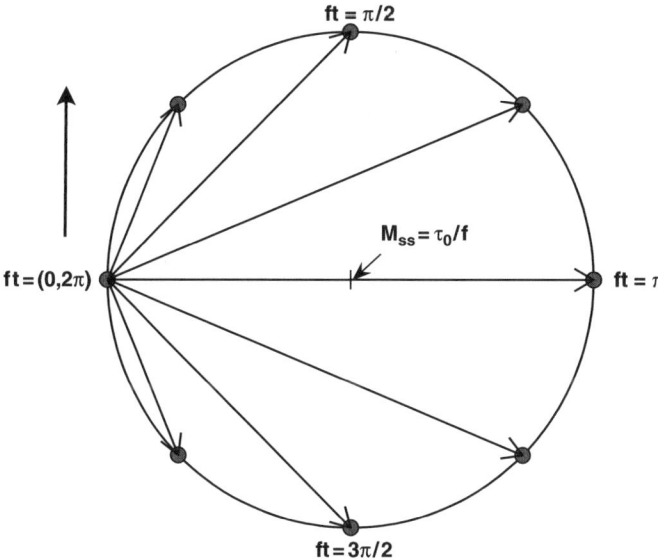

Fig. 2.1 Sketch of the solution to (2–22) for $\tau_0 = i\tau_0$. The circle repeats for each inertial period $t_p = 2\pi/f$

The advent of practical satellite navigation showed that inertial oscillations superimposed as cycloidal loops in the trajectories of wind-driven ice drift are ubiquitous, especially during summer when ice is relatively stress free and mixed layers tend to be shallow. An example from an unmanned buoy initially deployed near the North Pole in 2002 (Fig. 2.2) shows a well behaved series of inertial oscillations excited by a rapid increase in wind and drift velocity during the last half of day 269. By integrating a simple combination of mean velocity with a clockwise circular rotation from a starting point at time 270.0, the trajectory over the next two days can be reproduced reasonably well. The mean ice velocity is dominated by shear between the ice cover and the upper mixed layer in the direction of surface stress (Chapter 3), so in fact the actual velocity in the mixed layer was probably not much different from the highly idealized situation depicted in Fig. 2.1, at least for about four inertial periods.

A more complicated drift observed near the end of the SHEBA project (Fig. 2.3) illustrates a technique called complex demodulation applied to sea ice drift (McPhee 1988). The procedure uses least-squares error minimization to fit a differentiable function to observed positions over a suitable time interval, in order to isolate the inertial and diurnal tidal components of drift velocity. It also provides a rational estimate of drift velocity, when navigation fixes are not evenly spaced in time.

Expanding on concepts introduced by Perkins (1970), the drift velocity over a time interval comparable to the inertial and/or diurnal tidal period, is expressed as the sum of a mean part (V_0), plus oscillations from a combination of clockwise and counterclockwise rotating components:

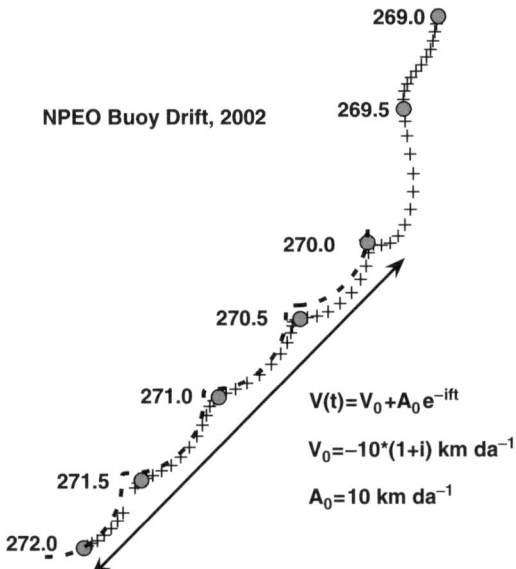

Fig. 2.2 Unmanned buoy trajectory from 26 to 29 September 2002. Plus symbols show satellite navigation fixes every half an hour. The dashed curve shows the integral of the simple velocity expression from the initial position at time 270.0

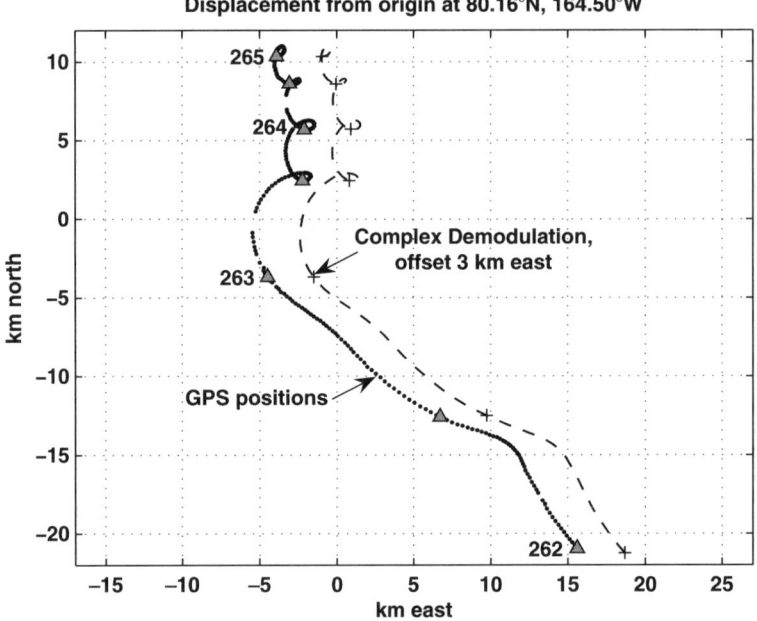

Fig. 2.3 Black dots show Global Positioning Satellite navigation fixes for the SHEBA drift station during a three-day period in September 1998. The dashed curve is position derived from complex demodulation, offset 3 km eastward for comparison

2.5 Inertial Oscillations

$$V(t) = V_0 + S_{cw}e^{-ift} + S_{ccw}e^{ift} + D_{cw}e^{-i\omega t} + D_{ccw}e^{i\omega t} \tag{2.23}$$

where ω is diurnal tidal frequency. A complex function for position is obtained by integration:

$$X(t) = X_0 + V_0 t + (i/f)\left[S_{cw}\left(e^{-ift} - 1\right) + S_{ccw}\left(1 - e^{ift}\right)\right] \tag{2.24}$$
$$+ \cdots (i/\omega)\left[D_{cw}\left(e^{-i\omega t} - 1\right) + D_{ccw}\left(1 - e^{i\omega t}\right)\right]$$

Other natural frequencies may be considered as well, but at high latitudes over relatively short time intervals, the frequency separation between inertial motion and semidiurnal tides is too small to have much impact on the computed velocity fields.

Given a time series of position fixes over a time period comparable to the longest period in the frequency array (24 h), solving for the coefficients in (2.24) becomes a linear algebra problem using standard least-squares analysis and Gaussian elimination (McPhee 1988). In a typical application, the vector of complex coefficients ($[X_0, V_0, S_{cw}, S_{ccw}, D_{cw}, D_{ccw}]$) is calculated every 3 h over a window 24 h wide, with velocity then calculated at any particular time via (2.23) by linear interpolation of the coefficients. A complex-demodulation trajectory calculated from the position data in Fig. 2.3 is drawn offset for comparison. An expanded view of the velocity field during September 1998 (Fig. 2.4) shows that in general, the inertial component is generally larger than diurnal except during times of rapid acceleration (change in inertia).

Before "undithered" global positioning satellite capability, estimating ice velocity from navigation data was hampered by relatively sparse position data or by relatively large errors in individual fixes. By complex demodulation, it was possible to obtain realistic velocity estimates from such data, because the technique incorporates the physical constraint of the inherent inertia in the coupled ice/upper ocean system. With the advent of frequent, highly accurate GPS data, other techniques provide good velocity data; nevertheless, complex demodulation offers considerable insight into the physical system. Relatively sudden accelerations or decelerations in ice drift (e.g., day 263) can set off persistent trains of inertial oscillations, with strength depending not only on wind (drift) speed but changes in direction with respect to the existing inertia of the ice/upper ocean system. Borrowing from electrical engineering terminology, the oscillating coefficients are like "phasors" that describe the amplitude and phase of the inertial or tidal oscillation. Figure 2.5 shows a time-series vector representation of the two leading coefficients in (2.23). For the first part of the period there is rapid drift to the northwest with relatively small inertial content. Then beginning about midday (UT) on day 262, the mean motion veers rapidly northward, and apparently this clockwise "kick" excites a strong train of inertial phasors that persists with little phase change (change in vector orientation) for several inertial periods with relatively small mean velocity, hence the pronounced cycloidal motion apparent in the drift trajectory (Fig. 2.3).

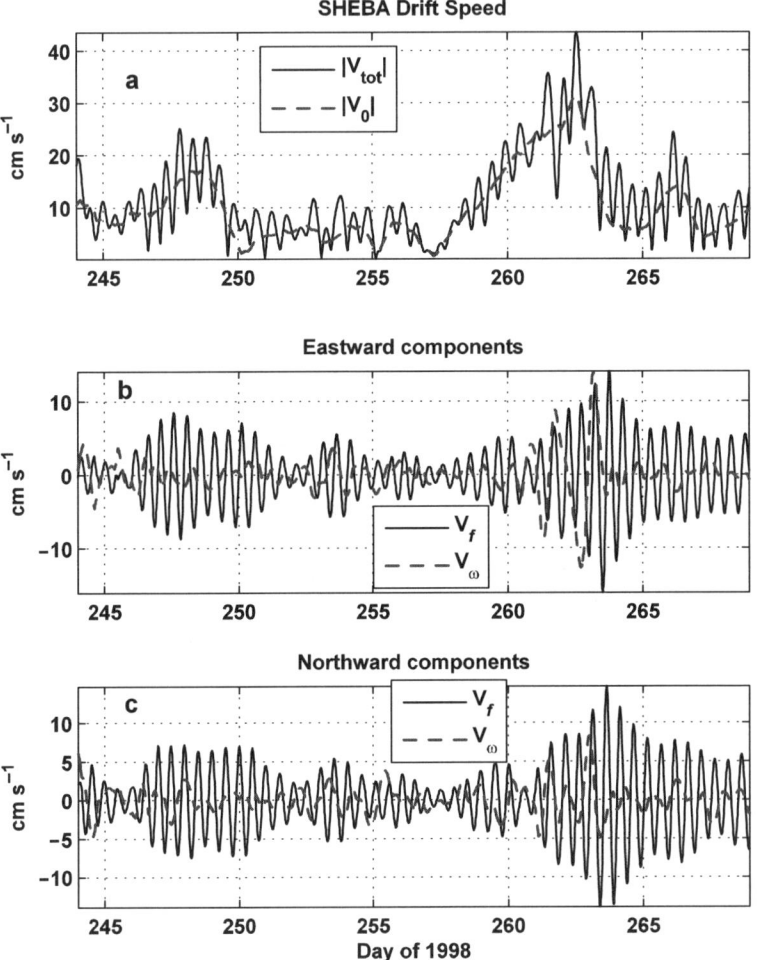

Fig. 2.4 a SHEBA station drift speed in September 1998. Heavy grey curve is The "mean" velocity, $|V_0|$ (heavy grey curve) was interpolated from complex demodulation fits every 3 h over a 24-h window. The heavy bar indicates the period shown in Fig. 2.3. **b** Eastward oscillating inertial and diurnal components. **c** Northward oscillating components

2.6 Ekman Pumping

To conserve mass, gradients in horizontal volume transport are accompanied by vertical motion. Neglecting the material derivative, the divergence of (2.21) is

$$\nabla \cdot M = -\frac{1}{f}\nabla \cdot (k \times \tau_0) = \frac{1}{f}\nabla \times \tau_0 \tag{2.25}$$

2.6 Ekman Pumping

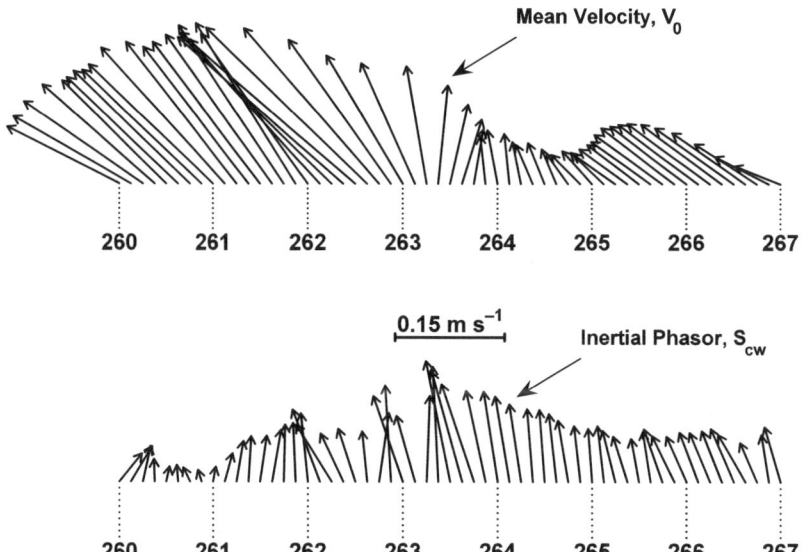

Fig. 2.5 Time series of the leading coefficients of the complex demodulation fitting for seven days in September, 1998. For V_0, up is north. The inertial phasor represents the amplitude and phase of the clockwise rotating inertial velocity. The initial orientation of S_{cw} is arbitrary

From the continuity condition

$$\nabla \cdot \boldsymbol{M} = -\int_{z_{bl}}^{0} \frac{\partial w}{\partial z} dz$$

To an observer on the ice, the vertical velocity at the bottom of the boundary layer is then proportional to the curl of the surface stress

$$w_{pyc} = \frac{1}{f} \nabla \times \tau_0 \quad (2.26)$$

In March 1998, at the SHEBA project in the central Arctic, we observed an upwelling event apparently related to intense local ice deformation (McPhee et al. 2005). Late in the afternoon (local time, UT—9 h) of March 19, the ice floe on the starboard of the ship, where most of the instrument systems were deployed, shifted forward by several hundred meters.[2] During this time, we measured turbulent heat flux in the upper part of the boundary layer that was at least an order of magnitude greater than at any other time during the year-long deployment; enough to cause significant basal melting when the ice is normally at its coldest. The particular event that occurred on March 19 (day 78 of 1998) was part of a more widespread period of

---

[2] A time-lapse video of the ice shear taken from the ship's bridge was shown on the CBS Nightly News.

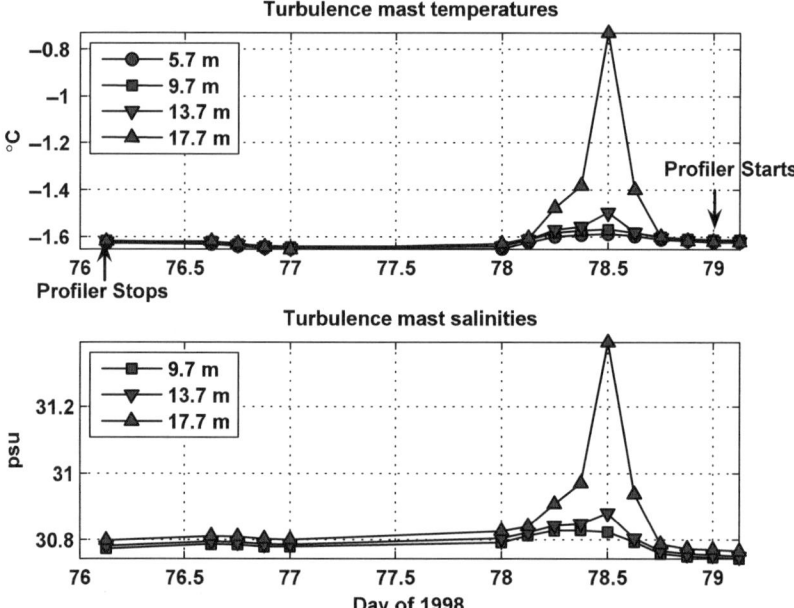

Fig. 2.6 Time series of temperature and salinity measured at the levels shown during March 1998. On day 78, the two lower clusters were enveloped in the upwelled pycnocline for several hours. During this time, turbulent heat flux at 9.7 m reached nearly 400 W m^{-2} (3-h average). During the time between the arrows, the automated profiler system was without power (Adapted from McPhee et al. 2005. With permission American Geophysical Union)

ice deformation, which played havoc with our oceanographic data gathering, mainly by severing the power line from the ship when a lead formed off the starboard side late on day 75.[3] Without ship power, the SHEBA automated profiling CTD system ceased operation early on day 76, and did not resume profiling until early on day 79. In the meantime, the turbulence mast was rigged to run on portable generator with sporadic coverage until it resumed full operation early on day 78 (UT). Records of temperature and salinity at constant levels in the upper ocean (Fig. 2.6) show well mixed conditions in the upper 18 m of the water column before and after day 78, but during that day there were large excursions indicating upwelling of pycnocline water. Figure 2.7 shows that the potential density measured at noon (UT) by the TIC at 17.7 m was about 0.5 kg m^{-3} larger than at the beginning and end of the day. Also shown are profiles of σ_0 from the automated profiler system at times 76.25 and 79.0, taken as representative of the ambient conditions surrounding the event. The maximum density observed at 17.7 m corresponds to the density at about 30 m in the surrounding undisturbed ocean. This brought water that was usually well below the active turbulence zone close enough to the surface that mixing was intense.

---

[3] The SHEBA experience in March 1998, accentuates an annoying, and to some extent unavoidable, aspect of ice camp measurements: that it is often difficult to keep the instrumentation operating in a rapidly changing environment, just when things get really interesting.

2.6 Ekman Pumping

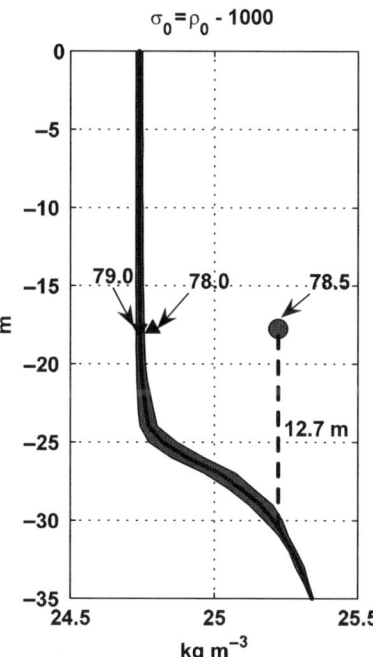

Fig. 2.7 Average σ_0 profile (solid) from the last 3 h before the power outage on day 76 and the first 3 h after power was restored on day 79 (limits indicated by shading). The symbols mark σ_0 values at the lowest cluster on the turbulence mast (17.7 m) at times indicated. The dashed line shows the difference in isopycnal elevation between conditions observed at time 78.5 and the undisturbed surrounding ocean

During the 3-h period centered at 1200 UT, measured turbulent heat flux was 150 and 460 W m^{-2} at 5.7 and 9.7 m, respectively. The latter was more than an order of magnitude greater than observed at any other time during the nearly year-long deployment.

The observed response was puzzling. There had been an episode of pressure ridge building about 110 m "upstream" of the turbulence mast just before the March 19 event, which could have conceivably caused a "wake" that would raise the pycnocline; however, we had not witnessed anything similar from flow across pressure ridges before (or after) during SHEBA. Application of a "large-eddy-simulation" model that would have resolved small enough scales to simulate the impact of a pressure ridge keel was unable to reproduce the observed upwelling (Skyllingstad et al. 2003). Since the increase in potential energy associated with the rise in the pycnocline was appreciable, it seemed unlikely that turbulent mixing alone could have accounted for the change, and as we drifted in response to the wind, the anomaly reverted to ambient in about the same amount of time that it formed.

A clue to the origin of the March 19 event came with post-project analysis of the synthetic aperture radar (SAR) imagery using the automated RGPS system (Kwok 1998). By tracking features in successive SAR images RGPS provided close to daily estimates of ice velocity and deformation on a uniform 5 km grid covering a domain about 200 km by 200 km, in the vicinity of the SHEBA drift station. An example of the velocity field product derived from two RGPS scenes spaced about one day apart is shown in Fig. 2.8. The colored contour map accentuates the boundaries between three comparatively uniform velocity zones, with the more southerly boundary passing near the ship trajectory from 77.75 to 78.73, i.e., during the time of the upwelling event.

Using Rossby-similarity (see Section 4.2.2) to relate ocean stress to ice velocity (assuming that geostrophic current was small in this region), we calculated a gridded map of kinematic stress corresponding to the velocities of the 5-km RGPS grid (Fig 2.8a). We then calculated a lower limit on the kinematic stress curl according to

$$\nabla \times \tau_0 \geq \frac{\Delta \tau_{0y}}{\Delta x} - \frac{\Delta \tau_{0x}}{\Delta y}$$

where the differentials are approximated by differences over the grid scale $\Delta x = \Delta y = 5$ km, with results shown in Fig. 2.8b. Although, the RGPS analysis cannot estimate shear (or stress curl) on scales smaller than 5 km, our observations from the ship and by analysis of SAR images in the vicinity of the ship suggested that the shear occurred across scales at least an order of magnitude less (i.e., 500 m). If the numerical value of stress curl from Fig. 2.8b, evaluated in the vicinity of the ship, is multiplied by 10, the resulting pycnocline displacement is about the same as observed (McPhee et al. 2005), and we thus inferred that the March 19 upwelling event was a result of Ekman pumping.

It is remarkable that the zones of intense stress curl, manifested locally at the ship by a dramatic shear across a narrow lead, extend at least 200 km in along more or less parallel arcs. The strength of the ice appears to provide a mechanism by which gradients in the forcing wind field are concentrated into narrow shear zones, which by Ekman pumping induce substantial isopyncnal displacement and much enhanced mixing of heat and salt in the upper ocean. The RGPS analysis reveals that these concentrated shear zones extend for long distances and are a ubiquitous feature in the Arctic ice pack (Kwok 2001).

2.7 The Equation of State for Seawater

In polar oceans, a layer of cold, less saline water nearly always overlies water that is both warmer and saltier. This negative temperature gradient by itself is destabilizing,[4] so stratification is maintained by the negative salinity gradient (i.e., increasing with depth). The equation of state for seawater is by convention expressed as a function of temperature, salinity as units of the practical salinity scale (abbreviated *psu*,

---

[4] In fresh water, thermal expansion changes sign at about 4°C, so fresh (or brackish water) may remain stable despite a negative temperature gradient near the surface. This does not hold for seawater with salinities in excess of about 24 units on the practical salinity scale.

2.7 The Equation of State for Seawater

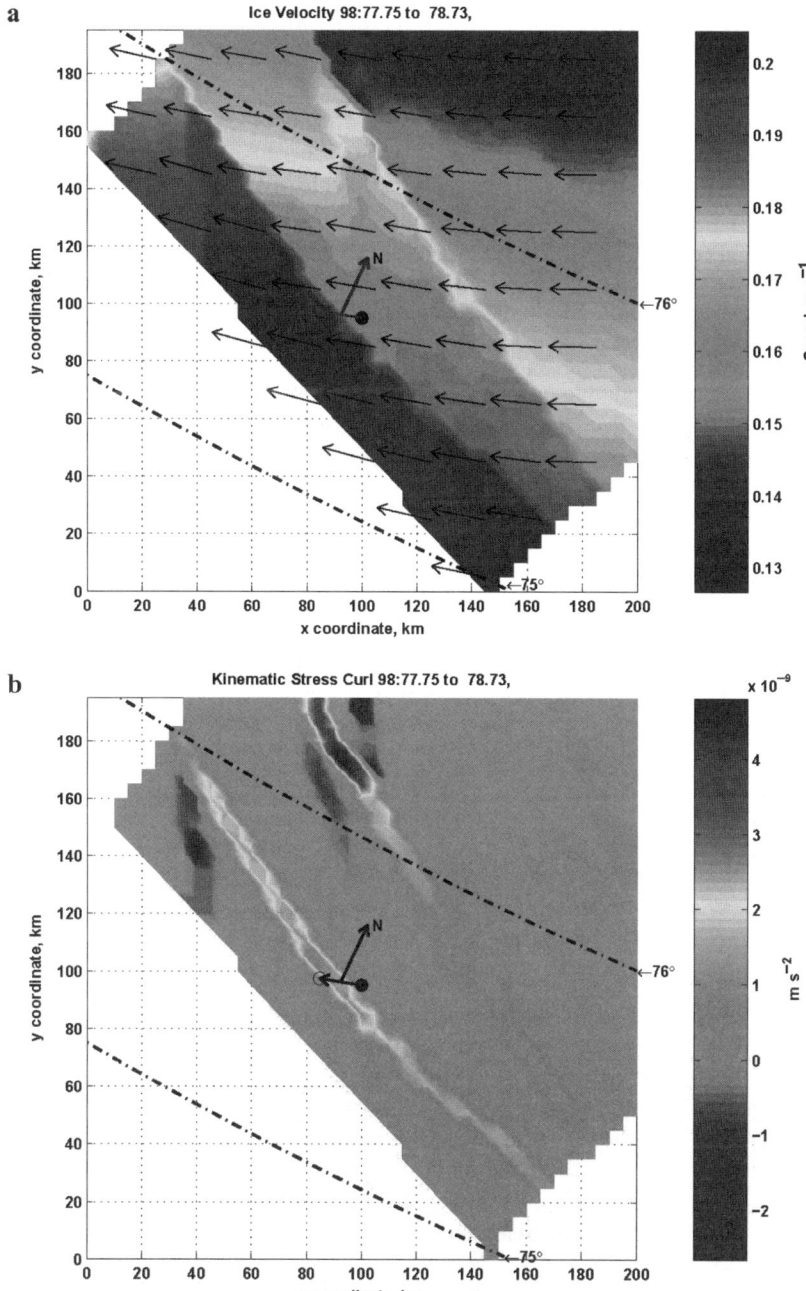

Fig. 2.8 a Average ice velocity field inferred from RGPS feature tracking of scenes separated by about a day. Circles indicated starting (solid) and ending (open) positions of the SHEBA drift station for the same period. Vectors are drawn every fourth grid point, one axis division is equivalent to $0.2\,\text{m s}^{-1}$. **b** Lower limit of kinematic surface stress curl, obtained by finite differences across the 5-km grid scale of the RGPS analysis (Adapted from McPhee et al. 2005. With permission American Geophysical Union)

corresponding closely but not exactly to the older expression *ppt*, parts per thousand), and pressure. Common oceanographic usage is to express pressure in terms of the departure from atmospheric pressure at the surface, with units of bars (10^5 Pa) or dbar (which corresponds reasonably closely with depth in m). The practical salinity scale relates the measured conductivity of seawater to an international standard, and thus provides a unique salinity for given conductivity, temperature, and pressure, all of which can be measured to high accuracy with modern oceanographic instrumentation. The UNESCO formulas for density as a function of the three state variables are given, e.g., by Gill (1982, Appendix 3) and are used in this work.

At low temperatures, the impact of changes in salinity on density is amplified relative to temperature changes because the thermal expansion factor, $\beta_T = -\frac{1}{\rho}\frac{\partial \rho}{\partial T}$, is small. The haline contraction factor ($\beta_S = \frac{1}{\rho}\frac{\partial \rho}{\partial S}$) is relatively insensitive to temperature, as illustrated in Fig. 2.9a, where variation of β_T and β_S with respect to their values at the freezing point are shown as functions of temperature. Over the range shown, β_T increases by over 400% while β_S remains within 2% of its freezing value. At constant pressure, the change in density may be expressed as

$$\frac{\delta \rho}{\rho} = \beta_S \delta S - \beta_T \delta T$$

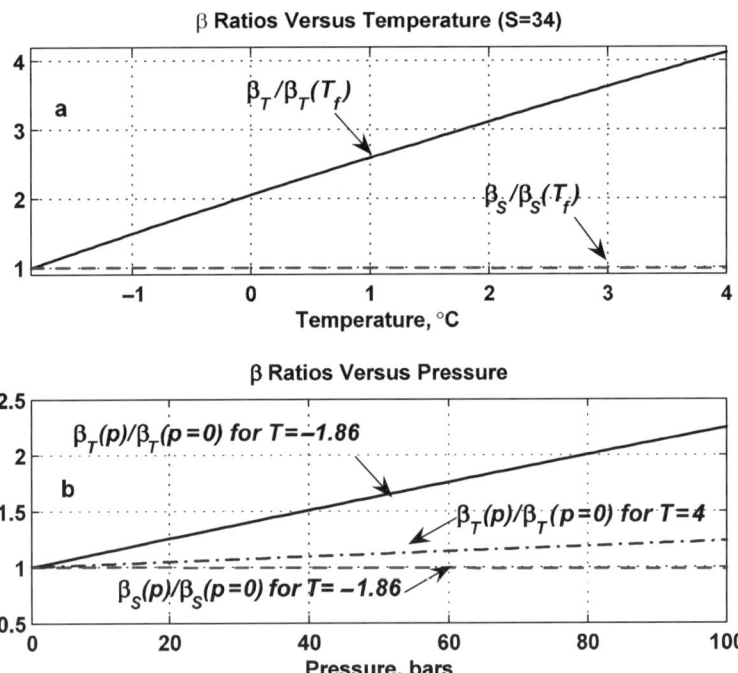

Fig. 2.9 a Ratio of expansion and contraction factors to their values for water at freezing temperature ($-1.86\,°C$) as a function of water temperature. At freezing the ratio β_S/β_T is about 33. At $T = 4\,°C$, it is about 8. **b** As in a, except ratios relative to the value at surface pressure (p = 0) as a function of pressure. At 400 m, the thermal expansion factor for water at freezing is about 1.5 times as large as at the surface. For water at $T = 4\,°C$, it is only about 1.1 times as large

2.7 The Equation of State for Seawater

To offset a change in salinity so that density remains unchanged would require that $\delta T/\delta S = -\beta_S/\beta_T$. For the conditions of Fig. 2.9a ($S = 34$ psu), this ratio is about 33 for $T = -1.86\,°C$ and about 8 for $T = 4\,°C$. Thus for water close to freezing, density variation is almost exclusively a function of salinity, and temperature may often be treated as a passive scalar contaminant.

In situ density depends on pressure, but in terms of the impact of vertical density gradient on dynamics, generally the pressure dependence is neglected by considering *potential density*, i.e., $\rho(T,S,p=0)$ or $\sigma_0 = \rho(T,S,p=0) - 1000$. The reason for this is clear: a well mixed layer with uniform T and S, will have a pressure induced vertical density gradient, but there is negligible work (besides friction) involved in moving a parcel from one level (pressure) to another. Yet there are idiosyncrasies associated with nonlinearities in the equation of state that make this less straightforward than it might at first appear. Consider, for example, the dependence of the expansion and contraction factors on pressure (Fig. 2.9b). Here the ratios of β_T and β_S to their values at surface pressure are plotted as functions of pressure. β_S has very little pressure dependence, but the magnitude of β_T increases with pressure. Plots are shown for two different temperatures to emphasize that the pressure dependence of β_T is much greater for cold water, resulting from the fact that cold water is more compressible than warm.

An example drawn from near Maud Rise in the Weddell Sea (Fig. 2.10) nicely illustrates certain consequences of nonlinearities inherent in the equation of state.

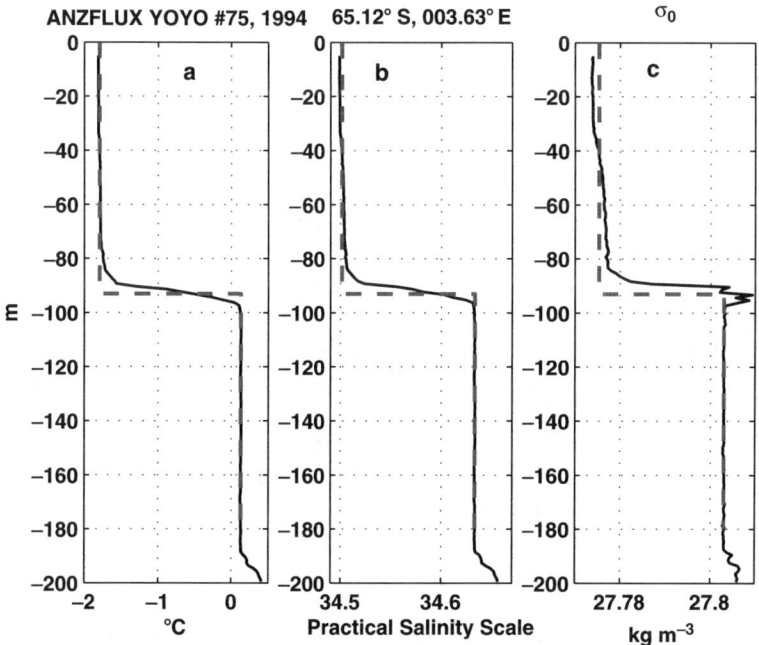

Fig. 2.10 YOYO Station 75 from ANZLUX 1994, on the eastern edge of Maud Rise in the Weddell Sea. **a** Temperature; **b** salinity; **c** σ_0 (potential density—1,000). Dashed lines are an idealized two-layer system based on the measurements

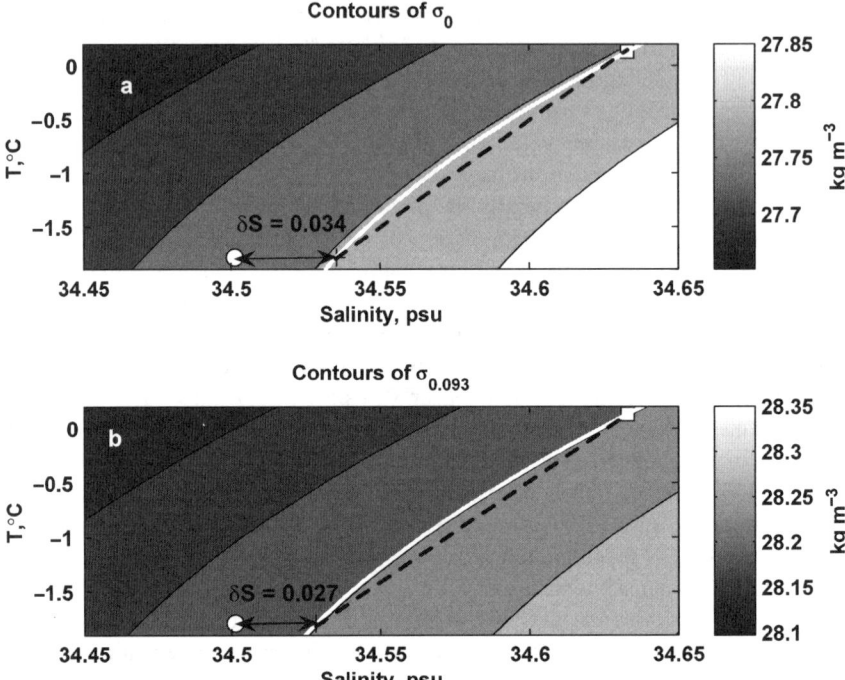

Fig. 2.11 Temperature/salinity diagrams with isopycnal contours for density calculated at **a** surface pressure and **b** at pressure corresponding to the mixed layer depth. *T/S* characteristics of the idealized two-layer system from Figs. 2.2 to 2.10 are indicated by symbols (circle for upper, square for lower). See text for further details (see also Plate 15 in the Colour Plate Section)

Measured T, S, and σ_0 profiles can be reasonably well represented in the upper 200 m of the water column by a two-layer system with an upper layer thickness of about 93 m. The potential density difference between the two layers is quite small, less than $0.03\,\mathrm{kg\,m^{-3}}$. In oceanography it is customary to compare water masses via a temperature-salinity diagram, as drawn in Fig. 2.11a. The *T/S* pairs representing characteristics of the two (idealized) layers from Fig. 2.10 are shown as symbols embedded in contours of σ_0. The σ_0 isopycnal passing through the *T/S* point for the lower layer (with $T_l = 0.13\,°C$, $S_l = 34.63\,\mathrm{psu}$) is shown in white. The double arrow indicates the increase in salinity needed to raise the potential density of the upper layer to that of the lower. All else being equal, the salt rejected from about 13 cm of additional ice growth (at the time the ice was about 35 cm thick) would accomplish this. The dashed line connecting the modified surface water and the deeper water in Fig. 2.11a is the so-called mixing line, which describes the *T/S* characteristics of any product from conservative mixing of the two different water masses. Because of the isopycnal curvature, the mixing line lies to the right of the isopycnal passing through both the deep water and modified surface water, so any mixture of the two water types is denser than either of the end members. Since no consideration of

2.7 The Equation of State for Seawater

pressure was involved in these arguments, the resulting instability arises from the dependence of β_T on temperature (Fig. 2.9a) and by convention is called *cabbeling*.

Figure 2.11b is like Fig. 2.11a, except that here the isopyncals are drawn for density evaluated at the pressure (depth) of the interface between the two layers, about 9.3 bar. At the higher pressure the slope of the isopycnals in T/S space is less than for surface pressure, which means that for a fixed salinity, the change in density associated with a given change in temperature is greater at depth. In this case, the upper layer only needs a salinity increase corresponding to about 10 cm of ice growth to reach the same *in situ* density as the lower layer. Thus instability will be triggered *before* the potential density of the upper layer reaches that of the lower layer. The term coined by McDougall (1987) for this pressure effect is *thermobaricity*.

A method for illustrating thermobaricity presented by Akitomo (1999) provides additional insight into the nonlinear equation of state issues, and is easily applied to the idealized two-layer system. Suppose that enough ice grows to increase the salinity of the upper layer by $\delta S = 0.027$ psu so that the *in situ* density of the two layers is the same at the interface, i.e.,

$$\rho(T_u, S_u, p_{93}) = \rho(T_l, S_l, p_{93})$$

The difference between the density of the two-layer upper ocean and an ocean with uniform T and S equal to the upper layer values:

$$\delta \rho = \rho(T, S, p) - \rho(T_u, S_u, p)$$

is plotted in Fig. 2.12. If a parcel of water from the upper layer (square marker) is displaced downward across the interface, it will be heavier than its ambient surroundings and will continue downward. A parcel displaced across the interface from below (circle) will be lighter than its surroundings and will continue to rise. Consequently thermobaricity is mechanism for enhanced mixing that draws from the potential energy of the destabilizing temperature gradient. Once started, the thermobaric process is self sustaining, and is probably an important component of mixing in marginally stable polar oceans like much of the Weddell in late winter. As indicated by Fig. 2.11, it is the curvature of the isopycnals in T/S space that leads to mixing driven by nonlinearities in the equation of state. To separate cabbeling and thermobaricity conceptually may be a question more of semantics than physics, but the important point is that whenever the temperature profile is in itself destabilizing, it is important to consider pressure effects.

If there are widespread regions in the Weddell where upper ocean structure is such that only a few decimeters of ice growth could trigger deep-reaching thermobaric instability, why does an ice cover exists there at all? Or put another way, why is the Weddell Polynya not a quasi-permanent feature? The answer apparently lies with what Martinson (1990) termed the "thermal barrier." Whenever heat is mixed up from below, it rapidly warms the mixed layer to the point where ocean heat flux to the ice undersurface exceeds conduction through the ice cover or loss from open water, and the ice begins melting. This introduces positive buoyancy that effectively

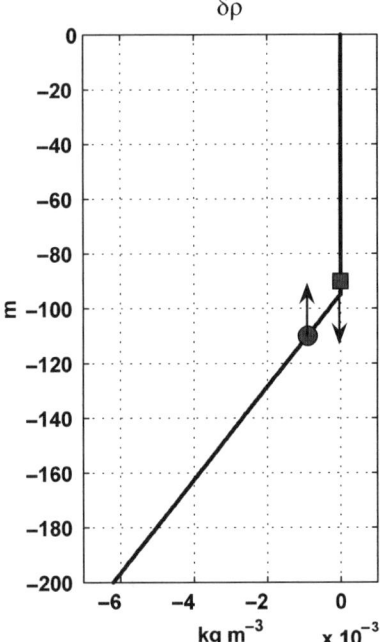

Fig. 2.12 Diagram of *in situ* density of the two-layer system minus the density of an upper ocean with uniform upper layer characteristics. Displacement of a water parcel with upper layer characteristics (square) downward makes it denser than its surrounding, while upward displacement from below the interface has the opposite tendency. The system is thermobarically unstable (see Akitomo 1999)

limits mixing driven from the surface, forming a new, shallower near surface layer. Thermobaric mixing below this layer may continue, driven by the nonlinearity, but is no longer affecting surface exchanges.[5]

The fact remains that if melting at the ice/ocean interface is too weak or too slow to counteract the combined effects of surface buoyancy loss from cooling and the cabbeling/thermobaricity mechanism at the base of the mixed layer, then convection will continue (McPhee 2003). Once the ice cover is gone and the air remains cold, there is nothing except horizontal advection of ice or fresh water to quell deep mixing, and essentially a direct connection between the abyssal ocean and the atmosphere is established. The Weddell Polynya demonstrated that such an event can have large, even global, impact.

A factor often ignored in the "mixing line" argument for instabilities arising from mixing of adjacent water masses with similar density but different T/S characteristics is that even in a fairly turbulent regime, diffusivities of heat and salt may differ. In Section 2.7 we described an upwelling event observed in March 1998, at the

---

[5] The conjecture that subsurface well mixed layers like that in the ANZFLUX profile in Fig. 2.10 between 100 and 180 m are remnants of mixing events where thermobaricity contributed is discussed further in Chapter 8.

2.7 The Equation of State for Seawater

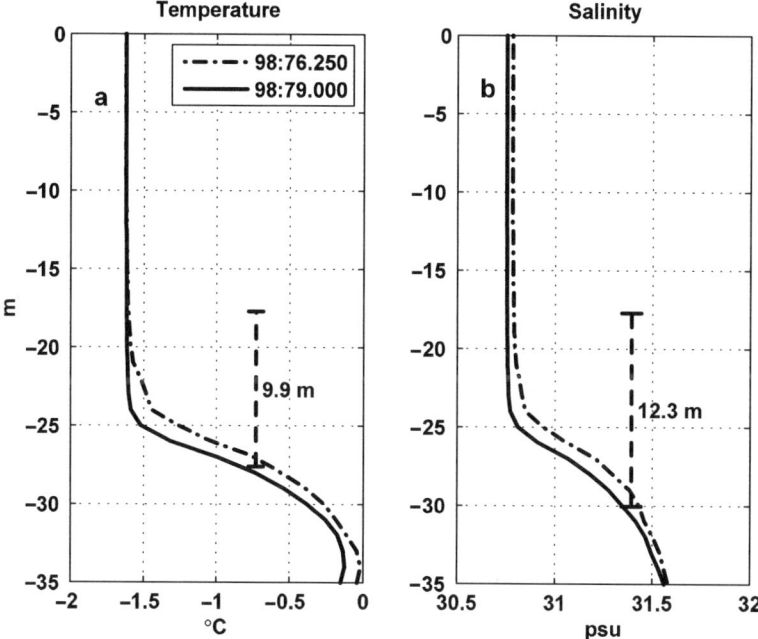

Fig. 2.13 SHEBA profiler temperature **a** and salinity **b** profiles bracketing the upwelling event on day 78. The dashed lines are the displacement of the mean isotherm **a** and isohaline **b** corresponding to measurements at 17.7 m on the turbulence mast at time 78.5

SHEBA station where isopycnals were observed to rise about 13 m above their ambient level in the undisturbed ocean, apparently in response to concentrated surface stress curl. Here we examine that event is more detail, as a possible example of double diffusion in a fully turbulent flow. Figure 2.13 shows the temperature and salinity profiles used to construct the σ_0 profile in Fig. 2.7. The upward displacement of the isotherm in the bracketing profiles to match the temperature observed on the bottom-most TIC at time 78.5 is indicated by the dashed line in Fig. 2.13a, and similarly for the matching isohaline in Fig. 2.13b. They differ by 2.4 m and both are less than the isopycnal displacement shown in Fig. 2.7. The logical explanation is that heat was mixed more efficiently than salt so that as the pycnocline fluid moved upward in response to Ekman pumping, its temperature lowered faster than its salinity. The loss in buoyancy from this additional cooling is why the isopyncnal displacement is about 0.4 m more than the isohaline displacement. In Fig. 2.14, the T/S properties of the water observed at 17.7 m at the maximum upwelling are compared with the ambient profiler T/S properties from all stations on day 79, averaged in 0.1 salinity bins. This strongly suggests that the upwelling event was capable of extracting heat from the upper pycnocline faster than salt.

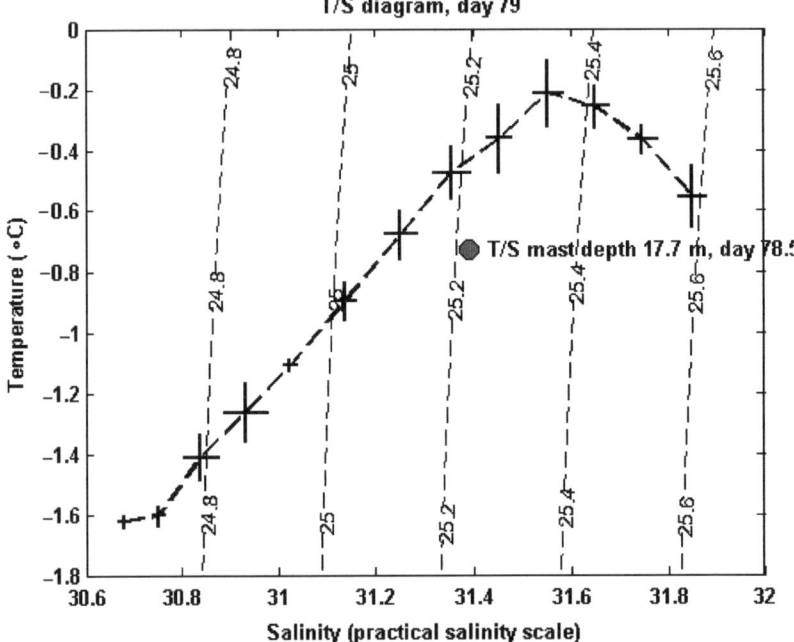

Fig. 2.14 Temperature/salinity diagram for all the SHEBA profiler casts on day 79, averaged in 0.1 psu bins. Sizes of the crosses correspond to twice the standard deviations of both T and S in each bin. The circle marks the T/S value for the 3-h average centered at time 78.5 at 17.7 m

References

Akitomo, K.: Open-ocean deep convection due to thermobaricity. 1. Scaling argument. J. Geophys. Res., 104 (C3), 5235–5249 (1999)

Ekman, V. W.: On the influence of the earth's rotation on ocean currents. Ark. Mat. Astr. Fys., 2, 1–52 (1905)

Gill, A. E.: Atmosphere-Ocean Dynamics. Academic, New York (1982)

Kwok, R.: The RADARSAT Geophysical Processor System. In: Tsatsoulis, C. and Kwok, R. (eds.) Analysis of SAR data of the Polar Oceans: Recent Advances, pp. 235–257. Springer, New York (1998)

Kwok, R.: Deformation of the Arctic Ocean sea ice cover: November 1996 through April 1997. In: Dempsey, J. and Shen, H. H. (eds.) Scaling Laws in Ice Mechanics and Dynamics, pp. 315–323. Kluwer, Dordrecht (2001)

Martinson, D. G.: Evolution of the Southern Ocean winter mixed layer and sea ice: Open ocean deepwater formation and ventilation. J. Geophys. Res., 95, 11641–11654 (1990)

McPhee, M. G.: Analysis and prediction of short term ice drift. Transactions of the ASME, J. Offshore Mech. Arctic Eng., 110, 94–100 (1988)

McPhee, M. G.: Is thermobaricity a major factor in Southern Ocean ventilation? Antarctic Sci., 15 (1), 153–160 (2003)

McPhee, M. G., Kwok, R., Robins, R., and Coon, M.: Upwelling of Arctic pycnocline associated with shear motion of sea ice. Geophys. Res. Lett., 32, L10616 (2005), doi:10.1029/2004GL021819

Pedlosky, J.: Geophysical Fluid Dynamics, Second Edition. Springer, New York (1987)

Perkins, H.: Ph.D. thesis: Inertial Oscillations in the Mediterranean. MIT/WHOI (1970)

Skyllingstad, E. D., Paulson, C. A., Pegau, W. S., McPhee, M. G., and Stanton, T.: Effects of keels on ice bottom turbulence exchange. J. Geophys. Res., 108 (C12), 3372 (2003), doi: 10.1029/2002JC001488

McDougall, T. J.: Thermobaricity, cabbeling, and water-mass conversion. J. Geophys. Res., 92 (C5), 5448–5464 (1987)

Tennekes, H. and Lumley, J. L.: A First Course in Turbulence. MIT, Cambridge, MA (1972)

Nomenclature

γ	Arbitrary fluid property
F^γ	Flux of γ
Q^γ	Source of γ
T	Temperature
S	Salinity expressed psu (units of the practical salinity scale)
ρ	Fluid density
p	Pressure
$\boldsymbol{R} = x\boldsymbol{i} + y\boldsymbol{j} + z\boldsymbol{k}$	Position vector
$\boldsymbol{u} = u\boldsymbol{i} + v\boldsymbol{j} + w\boldsymbol{k}$	Vector velocity
$\underset{\sim}{\tau}$	Reynolds stress tensor
$\boldsymbol{\tau} = \tau_{13}\boldsymbol{i} + \tau_{13}\boldsymbol{j} = \langle u'w' \rangle + i^* \langle v'w' \rangle$	Horizontal Reynolds traction vector
ω	Angular velocity
f	Coriolis parameter
M	Volume transport
$\mathbf{S_{cw}}, \mathbf{S_{ccw}}$	Clockwise and counterclockwise inertial components
$\mathbf{D_{cw}}, \mathbf{D_{ccw}}$	Clockwise and counterclockwise diurnal tidal components
β_T	Thermal expansion factor
β_S	Saline contraction factor
ρ_0	Potential density (density at surface pressure)

Chapter 3
Turbulence Basics

Abstract: When differential motion occurs between a sea ice cover and the upper ocean, momentum is exchanged across a turbulent boundary layer. If the heat and mass balance at the ice-ocean interface dictates ice growth or ablation, the turbulence will also transport heat and salt. This chapter introduces basic features of turbulence in natural flows, by describing general characteristics of turbulence; how it is measured in the somewhat unique under-ice environment; along with a discussion of how turbulent fluxes are estimated, including statistical significance and assumptions underlying the connections between time-series covariances and ensemble averages of turbulent fluctuation products. Simplified forms of the turbulent kinetic energy and scalar variance equations are described, and related to spectral characteristics including a length scale proportional to the inverse wave number at the peak in the vertical velocity spectrum.

3.1 General Characteristics

There exists little consensus on a precise definition of fluid dynamical turbulence. Hinze (1975) defines it thus: "Turbulent fluid motion is an irregular condition of flow in which the various quantities show a random variation with time and space coordinates, so that statistically distinct average values can be discerned." Tennekes and Lumley (1972) list pertinent characteristics of turbulent flow: (i) turbulence is irregular (as in Hinze's definition); (ii) it is highly diffusive, which causes rapid mixing and increases transfer rates; (iii) it occurs at high Reynolds number,[1] as instabilities from interaction of viscous and inertial forces manifest themselves; (iv) it is both highly rotational and three dimensional; (v) it is essentially dissipative, meaning that work must be done to maintain viscous losses to internal energy of the flow; and (vi) turbulence at high Reynolds number is a characteristic of the flow, rather than the particular fluid.

---

[1] $Re = UL/\nu$, where U and L are characteristic velocity and length scales in the flow and ν is kinematic molecular viscosity.

M. McPhee, *Air-Ice-Ocean Interaction*, 39–63.
© Springer Science + Business Media B.V., 2008

It is beyond the scope of this chapter to present anything beyond a cursory survey of turbulence. For the most part, where there has been substantial progress in turbulence theory, it has come from consideration of *fully developed* turbulence, i.e., turbulence that is homogeneous and isotropic, as in high Reynolds number flow some distance downstream from a wind-tunnel grid (e.g., Batchelor 1967; Frisch 1995). With very few exceptions, the IOBL flows considered here are essentially shear flows that vary with the strength of the wind or tide; are anisotropic at the scales of the energy-containing eddies; and, particularly under typically rough sea ice, are hardly homogeneous. Despite these shortcomings, we often find that a probabilistic description of the turbulent flows we measure in the IOBL provides a repeatable and useful tool for understanding turbulent transfers in boundary layers where rotation is important. By the same token it is important to keep in mind the limitations imposed by these departures from the assumptions often underlying turbulence theory. The primary goal of this chapter is to selectively investigate a few topics chosen from a vast field of turbulence research that are particularly germane to the IOBL problem. For a more thorough approach, the reader is referred to the texts referenced above.

3.2 IOBL Measurement Techniques and Examples

Compared with daunting technical difficulties faced in measuring turbulence near the surface of the open ocean (where orbital wave velocities and platform motion often dwarf turbulent fluctuations, except at very small scales), when working from sea ice, it is relatively easy to measure the covariance of vertical velocity and fluctuating horizontal velocity components that make up the horizontal Reynolds tangential stress. In many respects, pack ice forced by wind to drift over an otherwise nearly quiescent ocean provides a unique laboratory for studying boundary layer flow in a rotating reference frame. The ice itself is a platform that quells all but the longest period surface gravity waves, and allows us to suspend instruments at known depths through the entire extent of the boundary layer, moving at the maximum IOBL velocity. Additionally, at least away from obvious obstacles like deep pressure ridge keels, it presents a relatively uniform, flat surface, usually with comparatively small horizontal gradients in surface stress and upper ocean density.

3.2.1 Smith Rotors

During the 1972 AIDJEX Pilot Study in the Beaufort Gyre of the Canadian Basin, Prof. J. Dungan Smith of the University of Washington exploited the sea-ice laboratory by deploying arrays of small, partially ducted rotors, arranged in triads along orthogonal axes, providing for the first time, three-dimensional current velocities at several levels through an entire upper ocean boundary layer. In the face of conventional wisdom that held it to be impossible to measure turbulent fluxes directly in the ocean, Smith realized the potential for using underice measurements to address many unresolved questions about ocean mixing. In the 1972 experiment, his research team (including the author as a graduate student) deployed a total of 75 (25 triads) of the Smith rotors on three separate masts, two of which were placed in position by divers. The rotation rate of the rotors was sensed optically, with the period between electrical pulses triggered by mirrors on the rotor blades processed electronically and recorded on magnetic tape by one of the first available minicomputers (Data General Nova 1200 with an internal RAM of about 8,000 16-bit words—considered large at the time!).

Accurate measurement of vertical velocity is critical for covariance estimates of turbulent fluxes. Smith addressed the problem of resolving small vertical velocities with a mechanical current meter (sensitive to the angle of attack with respect to mean flow) by canting the "z-axis" of the current meter triad away from the vertical by $30°$, so that all three meters sensed a sizable fraction of the mean flow. Turbulence measurements from the suspended masts, along with results from a modern Guildline CTD profiler, resulted in a fairly comprehensive view of the turbulence structure in a nearly neutrally stratified IOBL (McPhee and Smith 1976).

3.2.2 Turbulence Instrument Clusters

In planning for the Marginal Ice Zone Experiments (MIZEX) north of Fram Strait in the Greenland Sea (McPhee 1983), we realized that turbulent heat flux from the ocean would be a major factor in the mass balance and survivability of sea ice encountering the open ocean with near surface temperatures well above freezing. At the time, direct measurement of turbulent heat flux via the covariance of vertical velocity with deviatory temperature (i.e., turbulent departures from mean fluid temperature) had not been made in the ocean. We approached Art Pedersen, founder of Sea-Bird Electronics, Inc. (SBE), who had recently combined a novel period-counting scheme with Wien-bridge circuitry in commercially available temperature (SBE 3) and conductivity (SBE 4) sensors, with the idea of incorporating a version of Smith's rotors (by then using a Hall-effect magnetic pickup in lieu of the earlier optical system) into a highly modified version of the SBE conductivity/temperature/depth (CTD) instrument. Pedersen responded positively, and assembled the SBE-35 system in his garage on Mercer Island, Washington, in time for use in the 1984 main MIZEX experiment in the Greenland Sea marginal ice zone. The SBE 35 could accommodate up to seven *turbulence instrument clusters* (TICs) each comprising three Smith rotors oriented along orthogonal axes, with the z-axis nominally $45°$ from vertical, mounted in the same horizontal plane as nearby SBE temperature and conductivity sensors. Five frequency signals (three low frequency output from the current meters and two high frequency from the T/C sensors) from each TIC were transferred by co-axial cable to the backplane of the SBE-35 deck unit, and recorded via computer. The new system provided credible estimates of ocean heat flux during the MIZEX experiment (McPhee et al. 1987), and substantially changed our view of how heat and salt are transferred at the ice/ocean interface, as described in Chapter 6.

The basic TIC configuration was modified slightly to reduce the diameter (hole size) and used in several experiments throughout the 1980s and 1990s, culminating in the year-long SHEBA project in 1997–1998. For CEAREX 1989, a new deployment scheme was developed in which TICs were mounted on a mast that could be lowered by winch to as much as 100 m below the surface. This entailed development of highly modified SBE 9 underwater units with input of 4–6 TICs, multiplexed through one sea cable to the surface deck unit. In this configuration the mast was oriented into an optimal angle of attack for the current meter triad by use of a vane.

For the 1992 LeadEX project, a further embellishment to the Smith-rotor TIC was addition of a SBE-7 fast-response microstructure conductivity sensor. The standard SBE 4 conductivity instrument uses a ducted design to increase accuracy and to maintain calibration, but the resulting flow constriction decreases the response to small turbulence fluctuations. Figure 3.1 shows the TIC configuration used during LeadEX and SHEBA. The exposed-electrode microstructure conductivity (μC) instrument is subject to sometimes severe calibration drift and conductivity spikes, but when used in combination with the nearby standard SBE 4, has provided credible direct estimates of salinity flux ($\langle w'S' \rangle$) and allowed us to calculate the total turbulent buoyancy flux from covariance measurements during LeadEX (McPhee 1994; McPhee and Stanton 1996).

Mechanical current meters are subject to several limitations, including biological fouling (occasionally a severe problem when ptinafora [jellyfish] are present during a short summer period). Possibly more significant is that although the individual

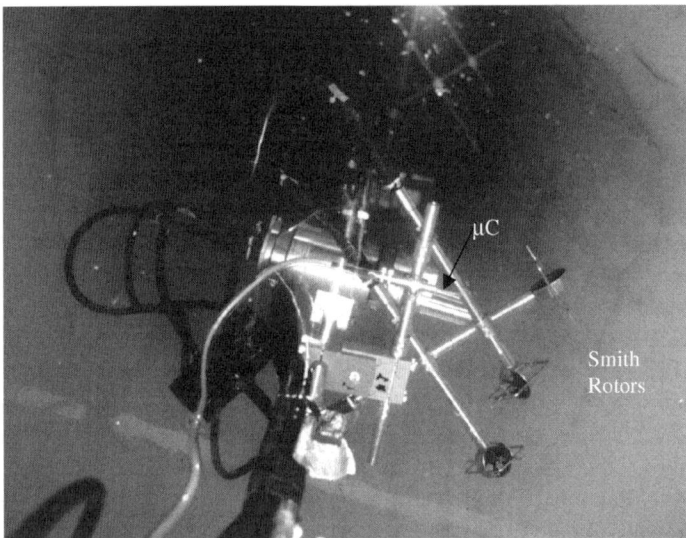

Fig. 3.1 Photograph of a "Smith-rotor" turbulence instrument cluster comprising three ducted rotors with Hall-effect magnetic sensors, along with SBE temperature (SBE 3), conductivity (SBE 4) and microstructure conductivity (SBE 7) instruments. Note that in this view, the upper current meter is horizontal while the other two are canted 45°

3.2 IOBL Measurement Techniques and Examples

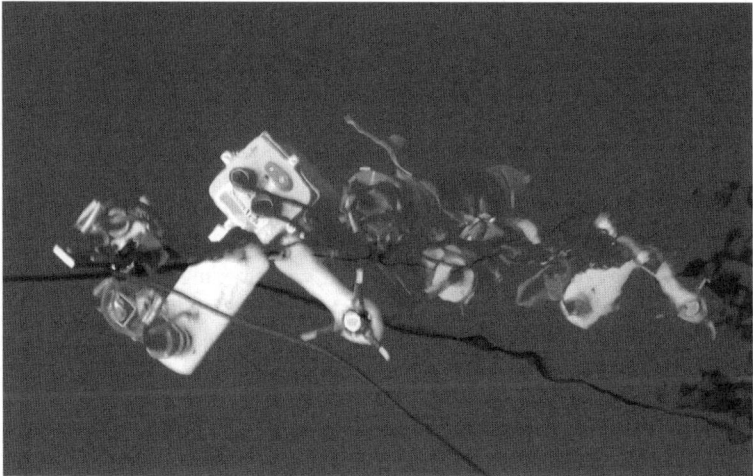

Fig. 3.2 Mast with two TICs equipped with SonTek 5 MHz ADVOcean current meters, near the surface during a deployment from the Baltic room of the R/V *Nathaniel B. Palmer* during the MaudNESS project, 2005

rotors will turn in currents down to about $1\,\mathrm{cm\,s^{-1}}$, when they are arranged along orthogonal axes, we have found that it takes a mean current of about $5\,\mathrm{cm\,s^{-1}}$ to keep all three rotors spinning, hence this imposes a practical minimum threshold for three-dimensional velocity measurement. Over the span of several experiments, we tried various other current measuring techniques including a laser-Doppler velocimeter (LDV) and a high-frequency (10 MHz) backscatter acoustic Doppler velocimeter (ADV). For the LDV and 10-MHz ADV instruments we found that a lack of optical and acoustic scatterers in the very clear polar mixed layers resulted in low signal-to-noise ratios, which severely limited their usefulness. However, during SHEBA, we tested a larger, 5-MHz version of the ADV (SonTek ADVOcean) that for the most part maintained adequate signal-to-noise ratios. Advantages of the ADV include a very low current threshold (determined again by the signal-to-noise ratio) and that it samples a volume (an ellipsoid with major axes about $1 \times 2\,\mathrm{cm}$) that is separated from the current meter apparatus by about 15 cm, significantly reducing flow disturbance. We exploited the low current threshold of the 5-MHz ADV to make credible estimates of momentum, heat, and salt flux in a gentle tidal current under fast ice in Van Mijen Fjord, Svalbard, in 2001 (McPhee et al. 2008, in press), and in several projects since, including a deployment from the Baltic room of the R/V *Nathaniel B. Palmer* during MaudNESS (Fig. 3.2).

3.2.3 Momentum and Scalar Flux Measurements

The main contribution of the TIC deployments has been to directly measure the covariance of vertical velocity with deviatory horizontal velocity, temperature, and salinity components. The Reynolds decomposition combined with Taylor's "frozen-field" hypothesis as described in Section 3.2.2 provide the primary rationale for estimating vertical fluxes from the covariance statistics. To illustrate this, we start with a 1-h time series of data, adapted from McPhee and Stanton (1996), taken during the 1992 LeadEX project in the Canadian Basin. Although this example is not typical of what we usually observe in the IOBL (others are presented below), it was chosen because it accentuates important aspects of the turbulent exchange process. At about the time of solar zenith on 7 April (year day 98), our TIC mast was positioned near the middle of the well mixed layer at the north edge of a kilometer-wide lead, which had opened the previous day and was freezing fairly rapidly with air temperatures in the -20 to $-28\,°C$ range. The ice and lead were drifting south at about $0.12\,m\,s^{-1}$, so that in our moving reference frame, the current appeared to come from the south, across the entire expanse the lead. The destabilizing buoyancy flux from salt rejected during freezing set up a boundary layer that appeared to be dominated by large scale eddies (perhaps "rolls") that advected past our instrument mast with some regularity. One of these "events" is indicated by shading of the vertical velocity (Fig. 3.3a) showing downward motion with a maximum approaching $3\,cm\,s^{-1}$, and persisting for nearly 5 min. During this time, temperature (b) and salinity (c) both deviate in a positive sense from their 1-h mean values. For salinity, this is consistent with salt ejected at the interface enhancing turbulence by gravitational acceleration. The large positive temperature anomaly indicates that, despite rapid freezing and upward heat conduction through the ice, water near the surface was being heated by incoming shortwave radiation that had penetrated the thin ice cover, and carried downward in the large eddies.

In the shaded sample and several other events, there is obvious correlation between w and the deviatory scalars, quantified by forming the product series, as shown for temperature and salinity in Fig. 3.3d and e, respectively. The product series are characterized by a number of significant upward and downward excursions, of roughly similar magnitudes, with peak values several times the 1-h mean values (indicated by the dashed lines). In many cases a large excursion of the product series is followed soon by an excursion of opposite sign, somewhat like one might expect in a wave field where the vertical velocity and displacement fields are $90°$ out of phase (the event between minutes 28 and 31 has somewhat this aspect). In such a field, the mean product over one or more periods is zero. This is perhaps one way to visualize how the inertial instabilities we term eddies work. Somewhat like a wave, most of the fluid carried up and down by the large-scale structures returns to its former level, without too much change in its properties. Nevertheless, during the large excursions there are smaller features (indicated by "fine structure" in the time series) that continuously "nibble" at the fluid within the large eddies, resulting in a relatively small net exchange of properties across the measurement level. This then shows up in the covariance (mean of the product series) provided the averaging time is adequate to capture a fair number of the large eddies. Averaged over the 1 h, the covariances suggest a downward heat flux of about $70\,W\,m^{-2}$ (roughly 15% the incoming shortwave radiation at the upper ice surface), and salinity flux comparable to that expected midway in the well mixed layer if the freezing rate was about 5 cm per day (McPhee and Stanton 1996).

3.2 IOBL Measurement Techniques and Examples 45

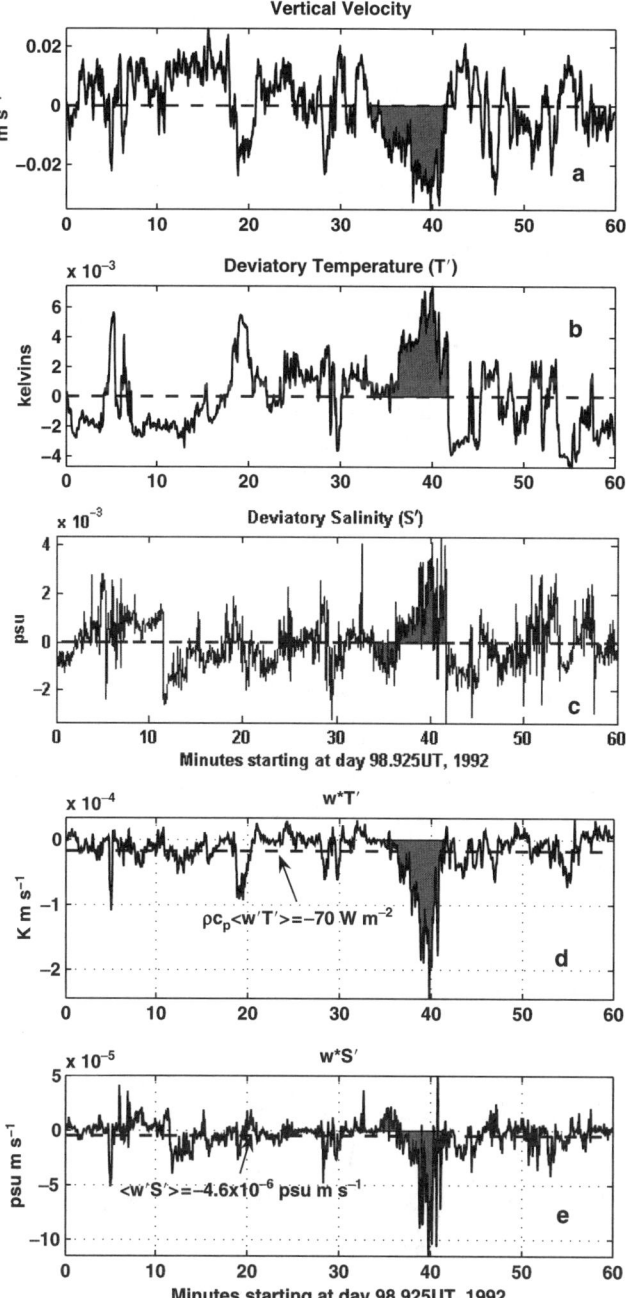

Fig. 3.3 One hour of turbulence data near midday at the edge of a freezing lead in April, 1992. **a** Vertical velocity at 14.8 m, approximately halfway through the well mixed layer. **b** Deviatory temperature. **c**. Deviatory salinity as measured with a microstructure conductivity sensor. **e** Product series of w times T', mean value is indicated by the dashed line, equivalent to a downward heat flux of 70 W m^{-2}. **e** wS' series, with downward salt flux from freezing. Shaded areas emphasize large downward eddy motion starting at about minute 35

3.2.4 Estimating Confidence Limits for Covariance Calculations

A variation on the bootstrap method (Efron and Gong 1983; Emery and Thomson 2001) provides estimates of confidence limits for the covariance statistics used to derive turbulent fluxes. The procedure is illustrated for the $\langle w'T'\rangle$ covariance from 1 h of turbulence data collected near the ice/ocean interface during one of the MaudNESS drift experiments. The data were separated into 15-min realizations, for which ADV velocities were rotated into a reference frame aligned with the mean streamline so that $\langle w \rangle = \langle v \rangle = 0$, and with a linear trend removed from temperature leaving the deviatory value T'. Time series for the first realization are shown in Fig. 3.4, including the product series $w \times T'$. Arrows indicate opposite conditions of downward velocity carrying a temperature deficit and upward velocity with a temperature surplus, each contributing positively to the instantaneous product time series. The average of the product deviatory time series (covariance) comprises both positive and negative contributions with fairly large excursions from the mean, which is 6.4×10^{-6} K m s^{-1}. With the sample mean so much smaller than the large-scale excursions (i.e., the standard deviation of the product time series is large relative to the mean), the question is: how representative of the true covariance is the mean of the product time series?

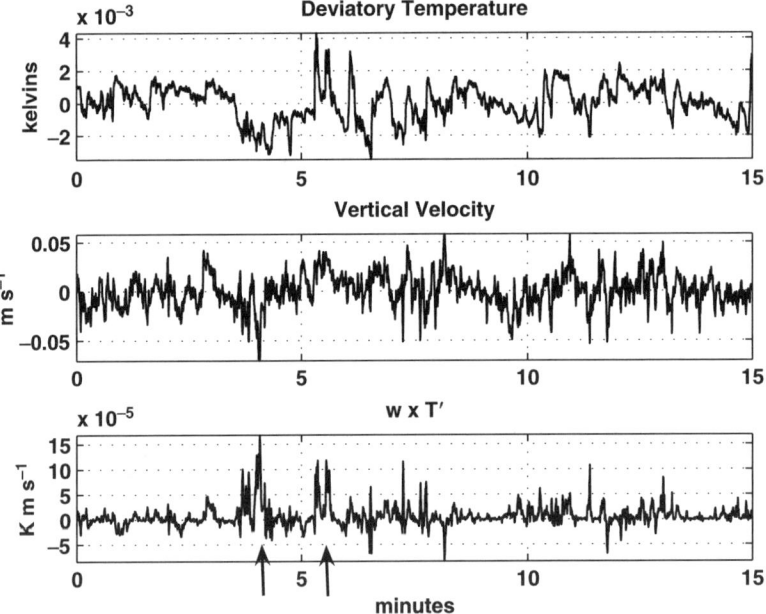

Fig. 3.4 Fifteen-min sample of T' and vertical velocity, and the product series centered at time 222.510 of 2005 during MaudNESS, Phase 2, Drift 1. Arrows mark times when downward velocity carrying a temperature deficit and upward velocity carrying a surplus both contribute positively to the covariance $\langle w'T'\rangle$

3.2 IOBL Measurement Techniques and Examples

With the bootstrap method, the time series is resampled randomly many times to build up a body of statistics for the likelihood of the sample mean occurring by chance. In practice this is done using a standard random number generator to produce a new artificial time series from the members of the original time series (not necessarily all, since one sample may reappear in the randomization many time), then calculating the mean of the artificial series. Done many times, this produces a new random variable, y, describing the means of the artificial series, from which a probability distribution function (pdf) may be estimated from the histogram of y values. In the present case, the expected value (true mean) of y is the $\langle w'T' \rangle$ covariance. To establish, e.g., upper and lower 95% confidence limits, abscissa values of the cumulative distribution function (cdf, integral of the pdf) were evaluated for ordinates 0.025 and 0.975, respectively. Figure 3.5 illustrates the procedure for the four 15-min turbulence realizations. In each case, 200 artificial time series of length 7,200 were synthesized by using a random number generator to assign indices from the original $w \times T_{data}$ time series (the upper right panel corresponds to the series shown in Fig. 3.4). Means of each artificial wT series were then assigned to the random variable y. Each panel in Fig. 3.5 shows 30-bin histograms of the y variable for each of the realizations in the hour-long data set, where the abscissa is

$$z = \frac{y - \bar{y}}{\sigma_y}$$

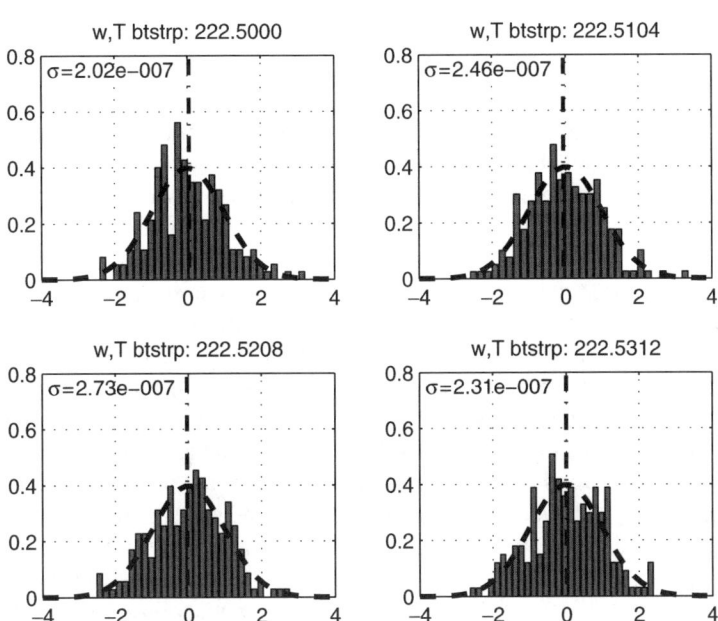

Fig. 3.5 Histograms of the artificial mean values (x) for 200 artificial time series from a random sampling of the original 15-min time series of the $w'T'$ products (length: 7200). Abscissa is $y = (x - \bar{x})/\sigma$ were σ is the sample std. deviation. Dashed curves are a normal distribution for zero mean and unit standard deviation

with σ_y the sample standard deviation of the y variable, and where the ordinate is

$$\frac{n}{N\Delta z}$$

where n is the number of samples in each bin of width Δz out of the total population N (equal to 200). The dashed curve in each case is the normal probability distribution function for a variable with zero mean and unit standard deviation:

$$p = \frac{1}{\sqrt{2\pi}} e^{-z^2/2}$$

After several examples comparing the bootstrap sample cdfs (obtained numerically) with the normal cdf showed very minor differences, we chose to modify the method slightly by assuming that y is normally distributed. Thus the 95% confidence interval is given by $-1.96 < z < 1.96$, or equivalently

$$CI = \bar{y} \mp 1.96\sigma_y$$

3.2.5 Averaging Time and the Spectral Gap

Heat flux values $(4 \times 10^6 \langle w'T' \rangle)$ with 95% confidence limits for each of the realizations (Fig. 3.6) differ by considerably more than their respective error bars. It is quite common to see large variation in covariance values from one 15-min sample to the next even in a flow that is relatively steady. The reason for this is implicit in the wT' time series of Fig. 3.4, where the covariance is dominated by a few large positive and negative events in which turbulent eddies with time scales of minutes transfer heat up and down. If a sample includes an excess or deficit of just a few of these events, it may affect covariance values appreciably. Thus sampling strategy represents a tradeoff between remaining in a "spectral gap" where the realization averaging time will capture most of the eddy events, but where changes in the mean flow speed and direction will not adversely affect covariance statistics. Generally, our approach has been to divide the flow regime into 15-min realizations for which the streamline coordinates adequately represent the actual flow, and then further average the covariance estimates for longer periods, typically 1–6 h. Using results from the bootstrap analysis for the individual realizations, covariance confidence intervals for the longer periods may be constructed by invoking the *central limit theorem* (e.g., Bowker and Lieberman 1959) where we assume that the covariances determined from each of the 15-min realizations are normally distributed with known variances. In that case, even with a small number of samples (in the present example, four values for $\langle w'T' \rangle$) the 95% confidence interval for the true mean, μ, is related to the sample mean by

$$CI_n = \bar{X}_n \mp 1.96\sigma_n/\sqrt{n}$$

3.2 IOBL Measurement Techniques and Examples

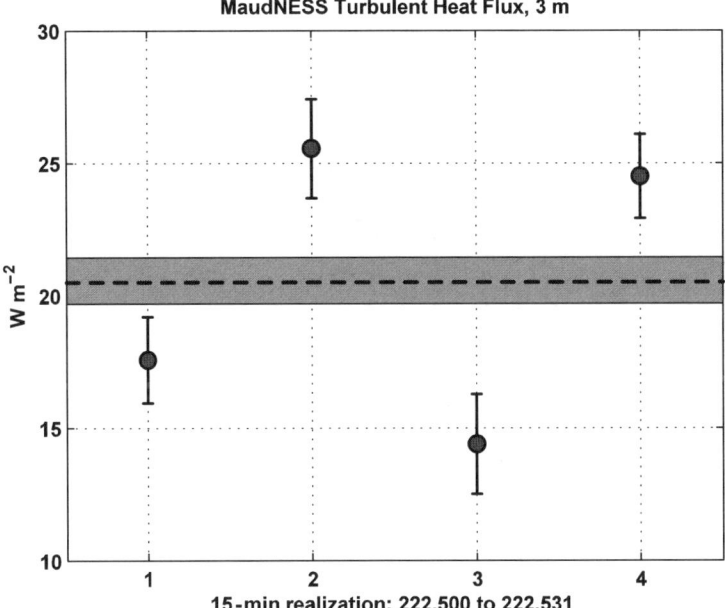

Fig. 3.6 Covariance samples from Fig. 3.5 multiplied by ρc_p to convert to sensible heat flux, along with confidence limits derived from the modified bootstrap method described in the text. The dashed line is the mean of four realizations (1 h) with the shaded box indicating the 95% confidence limits for the 1-h mean

where $\bar{X}_n$ is the sample mean of n realizations for the covariance, and σ_n is the average of the bootstrap standard deviations. The dashed line in Fig. 3.6 is the sample mean of the four covariance heat flux estimates, with the corresponding 95% confidence limits indicated by the shaded box.

The above procedure can be applied as well to evaluate confidence intervals for the covariance estimate of turbulent stress $\tau = \langle u'w' \rangle + i \langle v'w' \rangle$. In this case there are two dimensions, and the confidence limits trace a square in the complex plane, as depicted in Fig. 3.7.

Caution must be exercised in assigning a confidence interval for the covariance of two deviatory time series to the corresponding turbulent fluxes, mainly because of uncertainty in applying Taylor's hypothesis, which rigorously pertains only to steady flows. In practice, vary few natural IOBLs are steady for more than a few hours at a time, and the choice of averaging time for the "turbulent realizations" may significantly affect the mean fluxes estimated from the covariance measurements. An example from a 6-h period during the MaudNESS drift discussed above illustrates this. From 0300 to 0900 on 10 August 2005, ice drift was relatively steady with moderate stress. The data set was broken into 14 different sets of realizations of the product time series, $w \times T'$, with averaging times ranging from 1 min (360

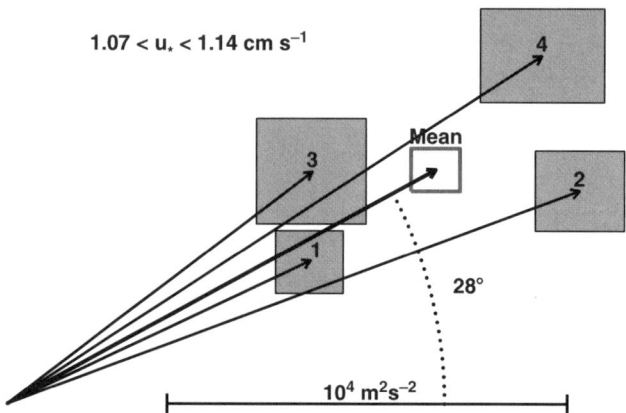

Fig. 3.7 Four 15-min realization covariance estimates of stress: $\tau = \langle u'w' \rangle + i \langle v'w' \rangle$ where shaded boxes show the individual confidence intervals for $\langle u'w' \rangle$ and $\langle v'w' \rangle$, respectively, along with the mean value and associated CI box. The real axis (horizontal) is aligned with the direction of mean flow in each case, with the average deflection of the stress vector about 28° counterclockwise. Also listed is a confidence interval for friction velocity magnitude, taken from the minimum and maximum limits of the mean stress

realizations) to 1 h (six realizations), and with the covariances, $\langle w'T' \rangle$, for each realization averaged for the entire period. Results with error bars representing the confidence intervals for the resulting mean covariances (multiplied by ρc_p) are shown in Fig. 3.8. When the realization interval is too short (<5 min) much of the covariance captured at the longer realization times is missed, because the interval is comparable to the time scales of the energy containing eddies. For sets with realization intervals in the range from about 6 to 20 min, the mean values lie within or near the confidence interval for the standard 15-min realization set (indicated by the dashed lines). Yet for longer realization intervals (30–60 min), the mean value is lower by as much as a watt per square meter.

This exercise is intended to drive home the point that calculating fluxes from covariance statistics of time series is only an approximation to the ensemble average and depends on several somewhat subjective factors, including a tradeoff between realization interval and longer term temporal changes in the environment. Based mainly on experience from numerous exercises like these, we typically use 15 min for the realization averaging interval, and further average these for a minimum of 1 h (four realizations) for stable flux estimates. A notable exception was LeadEX, where the scale of the energy containing turbulent eddies in the statically unstable conditions was significantly larger than in near neutrally stable conditions (McPhee and Stanton 1996). In that case the chosen realization interval was 1 h.

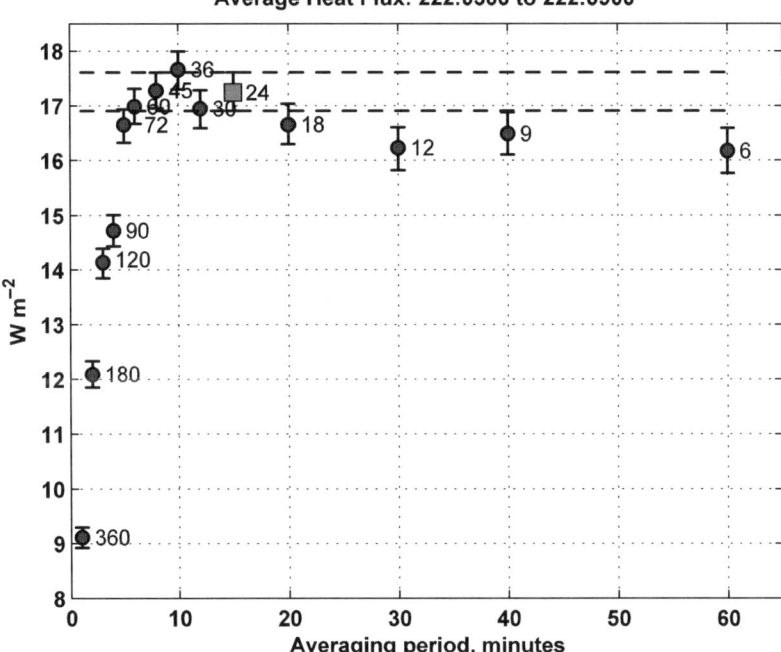

Fig. 3.8 Impact of averaging time on turbulent heat flux calculated from the covariance of deviatory temperature and vertical velocity for a 6-h period. Numbers next to the symbols indicate the number of realizations in each average. Error bars indicate the confidence interval for each average, derived via the bootstrap method as described in the text

3.3 Turbulent Kinetic Energy Equation

One of the most useful tools for understanding how turbulence works in natural flows is the *turbulent kinetic energy* (TKE) equation. Starting from the Boussinesq form of the Euler equation (2.16)

$$\frac{\partial \tilde{u}_i}{\partial t} + \tilde{u}_j \frac{\partial \tilde{u}_i}{\partial x_j} + 2\varepsilon_{ijk}\Omega_j \tilde{u}_k = -\frac{1}{\rho}\frac{\partial p}{\partial x_i} - \frac{g}{\rho}\rho'\delta_{i3} + \frac{\partial \tau_{ji}}{\partial x_j} \tag{3.1}$$

where $\tilde{u}_i = U_i + u_i$ is the instantaneous velocity, p is pressure after removal of the hydrostatic part, and Ω is the earth's rotation vector;[2] an equation for kinetic energy (per unit mass) of the flow is obtained by multiplying (3.1) by $\tilde{u}_i/2$. If the corresponding equation for the kinetic energy of the mean flow (U) is subtracted from the instantaneous equation, the result is (after eliminating some terms based on scaling

---

[2] In these equations, repeated indices imply summation as before, the Kronecker δ_{ij} is 1 for $i = j$ and 0 otherwise, and the alternating tensor is $\varepsilon_{123} = \varepsilon_{231} = \varepsilon_{312} = -\varepsilon_{321} = -\varepsilon_{213} = -\varepsilon_{132} = 1$ and 0 otherwise. Cartesian tensor notation is summarized by Hinze (1975, Appendix).

argument) the TKE equation (see, e.g., Hinze 1975; Tennekes and Lumley 1972):

$$\frac{\partial}{\partial t}\left\langle \frac{u'_i u'_i}{2} \right\rangle + U_j \frac{\partial}{\partial x_j}\left\langle \frac{u'_i u'_i}{2} \right\rangle + \langle u'_i u'_j \rangle \frac{\partial U_i}{\partial x_j} + \frac{1}{2}\frac{\partial}{\partial x_j}\langle u'_i u'_i u'_j \rangle \quad (3.2)$$

$$= -\frac{1}{\rho}\frac{\partial}{\partial x_i}\langle p' u'_i \rangle - \frac{g}{\rho}\langle \rho' u'_i \rangle \delta_{i3} - \nu \left\langle \left(\frac{\partial u'_i}{\partial x_j} + \frac{\partial u'_j}{\partial x_i}\right)\frac{\partial u'_j}{\partial x_i} \right\rangle$$

where $\langle u'_i u'_i \rangle$ (the trace of the Reynolds stress tensor) is twice the turbulent kinetic energy per unit mass.

Equation (3.2) is greatly simplified if the flow is steady and horizontally homogeneous (i.e., horizontal gradients negligible compared with vertical) with no mean vertical motion:

$$\underbrace{-\langle u'w' \rangle \frac{\partial U}{\partial z} - \langle v'w' \rangle \frac{\partial V}{\partial z}}_{P_S} \underbrace{- \frac{g}{\rho}\langle \rho'w' \rangle}_{P_b} = \underbrace{\frac{\partial}{\partial z}\left(\frac{\langle u'_i u'_i w' \rangle}{2} + \frac{\langle p'w' \rangle}{\rho}\right)}_{D} + \underbrace{\nu\left\langle \left(\frac{\partial u'_i}{\partial x_j} + \frac{\partial u'_j}{\partial x_i}\right)\frac{\partial u'_j}{\partial x_i}\right\rangle}_{\varepsilon}$$

(3.3)

In this case, the TKE balance comprises four terms. The first term on the left is the production rate of TKE by the interaction of turbulent stress with mean shear: $P_S = \tau \cdot \partial U/\partial z$. In shear flows as considered here, it is positive, meaning that the interaction of shear with Reynolds stress produces turbulence. A corresponding term with opposite sign occurs in the mean flow kinetic energy equation, which demonstrates mathematically how turbulence drains kinetic energy from the mean flow.

The second term on the left is the production of TKE by buoyancy (gravitational) forces in the fluid, and may be expressed in terms of *buoyancy flux*:

$$\langle w'b' \rangle = \frac{g}{\rho}\langle \rho'w' \rangle = -P_b \quad (3.4)$$

If TKE buoyancy production is positive (buoyancy flux negative), the flux of positive density variations is downward and energy is added to turbulence by gravity. If buoyancy production is negative (buoyancy flux positive), TKE is expended in moving denser fluid upward. P_b thus represents the conversion between TKE and potential energy of the fluid, and often has a profound impact on the scales of turbulence as explored in more detail in Chapter 5.

The term labeled D is the divergence of a combination of vertical flux of TKE and of the covariance of vertical velocity with turbulent pressure fluctuations in the flow. Both parts are difficult to measure accurately, and in the atmospheric surface layer (the first few dekameters), scaling arguments indicate that D is often negligibly small compared with the other three terms. For the outer part of the boundary layer, this is not necessarily valid, and indeed, long averages of TKE flux at multiple levels during the SHEBA project indicated the TKE flux divergence ($\partial \langle u'_i u'_i w' \rangle / \partial z$) to be important in the TKE balance (McPhee 2004).

The last term, ε, which involves gradients over small distances in all components of the flow, is the dissipation of TKE into internal energy (heat) of the fluid. It is where molecular viscosity plays a major role in natural turbulent flows. The simplest version of (3.3), with $P_b = D = 0$ (an approximate balance not uncommon in natural flows) results in a balance between production of TKE by shear and dissipation, and provides a framework for discussing the scales involved in transferring kinetic energy from large scale instabilities ("energy-containing eddies") where it is extracted from the mean flow, to scales where molecular interaction is important. Specific discussion of the energy-containing scales in IOBL turbulence is deferred to Chapter 5, but here we may anticipate that the shear production term can be expressed in terms of the friction velocity at the boundary, $u_* = \sqrt{\tau}$ where τ is the kinematic Reynolds stress magnitude, and a length scale characterizing the energy-containing containing eddies, λ:

$$P_S = \tau \cdot \partial U/\partial z = u_*^3/\lambda \quad (3.5)$$

Typical velocity and length scales in the IOBL are $u_* = 0.01 \text{ m s}^{-1}$ and $\lambda = 2 \text{ m}$, so a reasonable estimate of ε is $5 \times 10^{-7} \text{ m}^2 \text{ s}^{-3}$ (the units are equivalent to W kg^{-1}).

A fundamental concept in turbulence theory is that in flows with large Reynolds number, energy is "cascaded" from large (production) scales to scales small enough that inertial and viscous forces are comparable,[3] i.e., where the *local* Reynolds number of the flow is close to unity, $\upsilon\eta/\nu \sim 1$, υ and η being the small scale velocity and length scales, respectively. Kolmogorov hypothesized (see Batchelor 1967; Hinze 1975) that at these small scales, turbulence is statistically in equilibrium and uniquely determined by ε and ν. The velocity and length scales then follow from basic dimensional analysis (e.g., Barenblatt 1996)

$$\eta = \left(\frac{\nu^3}{\varepsilon}\right)^{1/4}$$
$$\upsilon = (\nu\varepsilon)^{1/4}$$

The viscosity of cold seawater is about $1.8 \times 10^{-6} \text{ m}^2 \text{ s}^{-1}$, so the Kolmogorov scales for our example are about $\upsilon = 1 \text{ mm s}^{-1}$ and $\eta = 2 \text{ mm}$. The scale velocities of eddies in the production and viscous subranges differ by a factor of about 10, while the length scales differ by three orders of magnitude. Thus there is a large range of length scales across which the "smaller whirls" feed. For comparison, if the fluid in our example were glycerine instead of water, with a kinematic viscosity of about $1.2 \times 10^{-3} \text{ m}^2 \text{ s}^{-1}$, the length scale would be $\eta = 24 \text{ cm}$, much closer to λ.

3.4 Scalar Variance Conservation

Conservation equations for scalar (T, S, or other contaminants) variance may be derived by analog with the TKE equation. The temperature variance equation for steady, horizontally homogeneous turbulent flow comprises a balance among

---

[3] The most concise description is still L. F. Richardson's famous verse: "Big whirls have smaller whirls that feed on their velocity, and little whirls have lesser whirls, and so on to viscosity."

three terms:

$$-\langle w'T'\rangle \frac{\partial T}{\partial z} = \frac{1}{2}\frac{\partial}{\partial z}\langle T'^2 w'\rangle + \nu_T \left\langle \frac{\partial T'}{\partial x_i}\frac{\partial T'}{\partial x_i}\right\rangle \quad (3.6)$$

where again the term on the left is the production of turbulent temperature variance (nearly always positive) balanced on the right by the gradient of a vertical transport term and "dissipation" of turbulent temperature fluctuations at molecular scales (ν_T is the kinematic heat diffusivity). Note that if the temperature gradient is known, and the thermal dissipation rate may be estimated (say, from microscale measurements of temperature fluctuations), then (3.6) provides an estimate of vertical heat flux, provided the transport term is negligible.

3.5 Turbulence Spectra and the Energy Cascade

Spectral analysis provides an important tool for studying the cascade of energy from large to small scales in turbulent flows. In many field situations, it is relatively easier to measure variance spectra compared with direct covariance between vertical velocity and the quantity in question. Spectral techniques also often serve as an important check on the validity of turbulence measurements. Several textbooks treat the subject well, both as it pertains to the turbulence theory (Batchelor 1967; Hinze 1975; Tennekes and Lumley 1972) and for methods of estimating spectra (the following draws heavily on techniques described by Bloomfield [1976]).

The one-sided spectrum is related to the variance of a quantity (in this case vertical velocity, w)

$$\int_0^\infty S_w(n)dn = \sigma_w^2 \quad (3.7)$$

where n is frequency (if measurements are made in the time domain, as is the most common case) and S_w is the spectral density of the time series w. Spectral density is defined as the Fourier transform of the autocorrelation function. In practice, it is estimated as follows. Given discrete samples of a deviatory time series ($x_k; k = 1\ldots N$) perform a discrete Fourier transform to obtain a vector X of length N, where

$$X(n) = \sum_{k=1}^{N} x_k e^{-2\pi i(k-1)(n-1)/N}; \quad 1 \le n \le N \quad (3.8)$$

The one-sided spectrum is first estimated by the *periodogram* calculated by

$$S_p(n) = \begin{cases} X_n \cdot X_n^*/N^2 & n = [1\ N/2+1] \\ 2X_n \cdot X_n^*/N^2 & n = [2\ldots N/2] \end{cases} \quad (3.9)$$

where the asterisk denotes the complex conjugate. The sum of all the elements of S_p is the variance of the x time series (from the discrete form of Parseval's

3.5 Turbulence Spectra and the Energy Cascade

theorem). If measurements are made in the time domain, the nth element of S_p in (3.9) corresponds to an estimate of the variance (or energy) in the signal at frequency $\omega_n = (n-1)/(N\Delta t)$ where Δt is the sample period. We use the same "frozen field" approximation as earlier to convert frequency to wave number, k, in a flow with mean velocity U past the sensor, namely $k = 2\pi\omega/U$. Note that this is the angular wave number, so that the wavelength in a spatially periodic flow with a peak at $k = 1\,\mathrm{m}^{-1}$ would be about 6.3 m.

The fundamental frequency interval is $\Delta\omega = \frac{1}{N\Delta t}$ so to satisfy the integral constraint of (3.7), requires that the frequency spectrum is

$$S' = S/\Delta\omega = N\Delta t S$$

and similarly the wave number spectrum is

$$S'' = S/\Delta k = \frac{N\Delta t U}{2\pi} S$$

Consequently the quantities $nS = \omega S' = kS''$ are invariant with the frequency or wave-number interpretation, hence are called the *area-preserving* form of the spectrum.

The periodogram of a finite time series differs from the true spectrum because it necessarily involves convolution of the spectrum with the transform of a rectangular window representing the sampling period. Many strategies exist for smoothing or otherwise manipulating the periodogram to accentuate salient features in the frequency (or wave number) domain. The less the time series is dominated by specific frequencies, the more variation is expected in particular spectral estimates from one realization to another (see, e.g., Jenkins and Watts 1968, Chapter 6). Turbulence is random enough to generally preclude sharp spectral peaks,[4] and we have found that smoothing of the periodogram with successive passes of a modified Daniell filter, following Bloomfield (1976), improves the estimates at lower wave numbers. We typically further average the estimates in equally spaced bins of $log_{10}(k)$ which greatly reduces variance at higher wave numbers, where many estimates occupy one bin. By aligning each individual spectrum (typically calculated from 15-min realizations of the flow as described above) onto a common $log_{10}(k)$ grid, it is possible to average spectra for several hours in a relatively steady flow, then fit with high-order polynomials, as shown in Fig. 3.9, from McPhee and Martinson (1994). Confidence limits for the actual w spectrum average are indicated by the shaded area.

In order to explore the utility of spectra like those shown in Fig. 3.9 requires discussion of the *inertial subrange* part of the energy cascade. We start with a simple dimensional argument. Suppose that somewhere between the big and tiny "whirls" of the production and dissipation scales, respectively, there is a range of scales for which the eddies are not "aware of" their larger and smaller cousins, and that the

---

[4] Indeed, if a sharp peak appears in the spectrum, it often signals some extraneous source besides turbulence. Examples are electronic or acoustic noise when the signal-to-noise ratio is small for acoustic current meters, or strumming of the mast on which they are mounted. Usually it is possible to filter these effects.

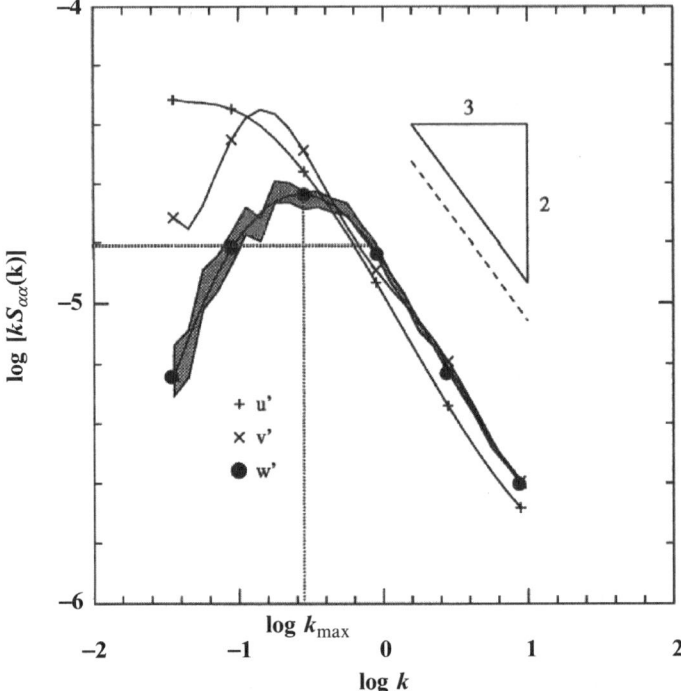

Fig. 3.9 Averaged (6 h) component velocity variance spectra from 20 m below the ice-water interface during a storm at Ice Station Weddell in 1992. Envelope indicates 95% confidence interval for the w spectrum only. The vertical line indicates the wave number at the maximum in the weighted (area-preserving) w spectrum used to estimate mixing length. The horizontal line identifies the spectral level used to estimate TKE dissipation (Reproduced from McPhee and Martinson 1994. With permission American Association for the Advancement of Science)

energy at those scales depends only on the local wave number and dissipation ε, i.e., the rate at which energy is being cascaded to the small scales. Following the notation of Barenblatt (1996), where square brackets denote dimensions, we have a dependent quantity (w spectral density) that depends on two governing parameters with independent dimensions.

$$[S_w] = L^3 T^{-2} \qquad \text{dependent quantity}$$

$$\text{\textemdash\textemdash\textemdash\textemdash\textemdash}$$

$$[k] = L^{-1} \qquad \text{governing parameters}$$
$$[\varepsilon] = L^2 T^{-3}$$

In this case, by the PI theorem a dimensionless quantity combining the dependent and governing parameters must equal a constant, i.e.,

3.5 Turbulence Spectra and the Energy Cascade

$$\frac{[S_w]}{[k]^{q_1}[\varepsilon]^{q_2}} = 0$$

which after solving for q_1 and q_2 yields

$$\frac{k^{5/3}S_w}{\varepsilon^{2/3}} = \text{constant}$$

Thus in that part of the area-preserving spectrum (Fig. 3.9) where the above assumptions hold, we would expect the spectra to fall off with a $-2/3$ slope (in the log-log representation), as indicated by the triangle. By convention, the equation for the vertical (cross-stream) spectrum is expressed as

$$S_w(k) = \frac{4\alpha_\varepsilon}{3}\varepsilon^{2/3}k^{-5/3} \tag{3.10}$$

where α_ε is the Kolmogorov constant for the along-stream spectrum, S_u. It is found from laboratory and atmospheric studies to have a value equal to about 0.51 when k is the angular wave number (e.g., Edson et al. 1991). The 4/3 factor in (3.10) comes from theoretical considerations of homogeneous, isotropic turbulence where it is possible to relate the one-dimensional along-stream and cross-stream spectra to the total (three-dimensional) energy spectrum (Batchelor 1967). In addition to deriving the "minus five-thirds" spectral shape for the inertial subrange in agreement with the dimensional analysis, the theory provides a relationship between the cross-stream (v,w) and along-stream (u) spectra (measured in the direction aligned with the mean velocity):

$$S_w = S_v = \frac{1}{2}\left(S_u - k\frac{\partial S_u}{\partial k}\right) \tag{3.11}$$

from which it follows that in the $-5/3$ slope region, the cross-stream spectra should be 4/3 the magnitude of the along-stream spectrum. We stressed above that IOBL shear flows are clearly anisotropic at large scales, and obviously from Fig. 3.9, S_w and S_v differ widely from each other and from S_u at small wave numbers. Yet at wave numbers around $3\,\text{m}^{-1}$ ($\log_{10}k \sim 0.5$), they are about equal, and are roughly 4/3 larger than S_u (indicated by the separate dashed line near the triangle), thus at these scales the turbulence appears to be more isotropic than not. The separation between along- and cross-stream spectra is often taken to mark the inertial subrange. Note that any point in the $-2/3$ range of the w spectrum in Fig. 3.9 suffices to estimate ε, e.g., the horizontal line intersects at the spectrum at $k=1$ with ordinate about $\log_{10}(kS_w) \approx -4.8$, which from (3.10) provides $\varepsilon \approx 1.1 \times 10^{-7}\,\text{W kg}^{-1}$.

If area-preserving velocity spectra exhibit the $-2/3$ log-log behavior, it lends confidence that we have made measurements at scales small enough to capture most of the covariance of the large-scale eddies. At scales where the turbulence approaches isotropy, the covariance contribution is very small. This can be confirmed by considering the *cross-spectrum* formed by smoothing, for example, the complex product of the w Fourier components and the complex conjugate of the T' components (Bloomfield 1976). The complex cross-spectrum is characterized by coherence (magnitude squared, normalized) and the phase angle, which is the arctangent of the imaginary part (quadrature spectrum) over the real part (cospectrum). In general, a wave number band will contribute to vertical flux if the cospectrum dominates (phase near 0 or π), while bands with phase near $\pm \pi/2$ (quadrature dominant) contribute little. Examples of $w\,T'$ and $w\,S'$ cross-spectra for the LeadEx measurements are given by McPhee and Stanton (1996).

3.6 Mixing Length, Eddy Viscosity, and the w Spectrum

In Section 3.3 we introduced a length scale of the energy-containing eddies, λ, and used it in the expression for shear production given by (3.5). Discussion of IOBL turbulence scales is the subject of Chapter 5, but here we anticipate those results by identifying λ with the wave number at the peak in the area-preserving w spectrum, $k_{\max}$, and introduce the concept of *eddy viscosity* in a turbulent flow. Eddy viscosity provides a conceptual method for closing the turbulence problem at "first-order" by relating fluxes of momentum and scalar properties in a shear flow to their respective gradients. By analogy with kinematic viscosity which depends on the products of velocity (internal energy) and mean free path of the molecules in the fluid, eddy viscosity is commonly represented by the product of a turbulent velocity scale and a length scale over which the dominant eddies in a flow are effective at diffusing momentum (this is different from the actual scale of the eddy motions):

$$K = u_\tau \lambda$$

In the simplest form of turbulent shear flow near a boundary where buoyancy and rotation are unimportant, the velocity profile is logarithmic with distance from the boundary, and the pertinent scale velocity is u_{*_0}, the square root of kinematic boundary stress. From this it follows immediately (Section 4.1) that $\lambda = \kappa z$ where κ is von Kármán's constant. If ε is known in this type of flow (say from measuring the spectrum), the magnitude of the turbulent stress is simply $\tau = (\kappa z \varepsilon)^{2/3}$. The problem for the IOBL, however, is that the linear dependence of λ on z is limited to at most a few meters from the boundary, and determining the scale of turbulence in the outer part of the boundary layer becomes a central issue in understanding turbulent transfer there. It appears that the inverse of the wave number at the peak in the w spectrum (but not the u and v spectra) provides a consistent estimate of λ, hence eddy viscosity. Measuring this at discrete levels through the entire IOBL then provides an important observational constraint on models that purport to simulate turbulent exchanges in the PBL.

Application of this concept to the IOBL dates from the 1972 AIDJEX Pilot Study data (McPhee and Smith 1976). We found that peaks in the area-preserving spectral

3.6 Mixing Length, Eddy Viscosity, and the w Spectrum

density of vertical velocity variance increased with depth for the first couple of turbulence clusters (to about 4 m from the ice) but that for greater depths (8–26 m), the peaks occurred at roughly the same wave number. We were also aware that Busch and Panofsky (1968) had shown that velocity variance spectra measured in the atmospheric boundary layer had the following characteristics: (i) wavelengths at the maximum in the logarithmic (area-preserving) spectra of vertical velocity increased linearly with height up to about 50 m, and more slowly beyond; (ii) the dimensionless frequency, $f_m = nz/V$ where n is frequency at the spectral maximum and V is mean wind speed, scaled with Monin-Obukhov similarity in the surface layer (discussed in Chapter 4); (iii) there was a relatively uniform shape to the normalized w spectra, when the abscissa values were scaled by f_m and ordinate values by u_{*0}^2; and (iv) the longitudinal (u) spectra did *not* show similarly predictable behavior. For the neutrally stratified surface layer, this means that the wave number at the maximum in the area-preserving w spectrum ($k_{\max} = f_m/z$) is inversely proportional to z hence to $\lambda_{sl} = \kappa z$. The fact that this dependence weakened for heights greater than about 50 m led Busch and Panofsky to speculate that the connection between λ and $k_{\max}$ might persist beyond the surface layer.

We reasoned (McPhee and Smith 1976) that if the increase in λ with distance from the boundary reached some limit comparable to κ times the surface layer thickness (a few meters in the IOBL), then our observations of $k_{\max}$ behavior at the various levels would be consistent with $\lambda = c_\lambda/k_{\max}$, as suggested by Busch and Panofsky. Experiments since (described in Chapter 5) have corroborated this view. The proportionality constant appears to be about 0.85. The vertical line in Fig. 3.9 thus suggests that the master turbulent length scale at 20 m during the 6-h event shown was about 3.2 m. If shear production and dissipation balance, (3.5) then implies that the Reynolds stress, $\tau = (\lambda\varepsilon)^{2/3}$, was about 5×10^{-4} m^2 s^{-2}, in agreement with direct measurements (see Fig. 5.3). Note that there is enough information from the w spectrum by itself to estimate the eddy viscosity at 20 m: $K \approx \varepsilon^{1/3}\lambda^{4/3}$, which is about 0.022 m^2 s^{-1}.

The isotropy condition indicated by the 4/3 separation between w and u spectra as shown by the ISW data (Fig. 3.9) is not always present in the IOBL measurements, particularly at levels closer to the surface. During the SHEBA project, for example, average normalized w spectra were remarkably similar at four levels ranging from 4 to 16 m from the ice undersurface (see Fig. 3.6 of McPhee 2004). However, for the two upper TICs, the u spectrum was consistently more energetic in the inertial subrange (as indicated by the w spectrum). We attributed this to a lack of horizontal homogeneity in the underice surface, as TKE advected from a prominent pressure ridge keel, often about 100 m "upstream" from the turbulence mast, spread vertically. In this case, we found that the gradient of TKE flux played a significant role in the TKE equation, and developed an alternative (but closely related) method for estimating the magnitude of stress from the w spectrum by considering the production rather than dissipation of TKE, resulting in a simple relation

$$u_*^2 = \frac{\phi}{c_\gamma}\gamma_*^{2/3} \tag{3.12}$$

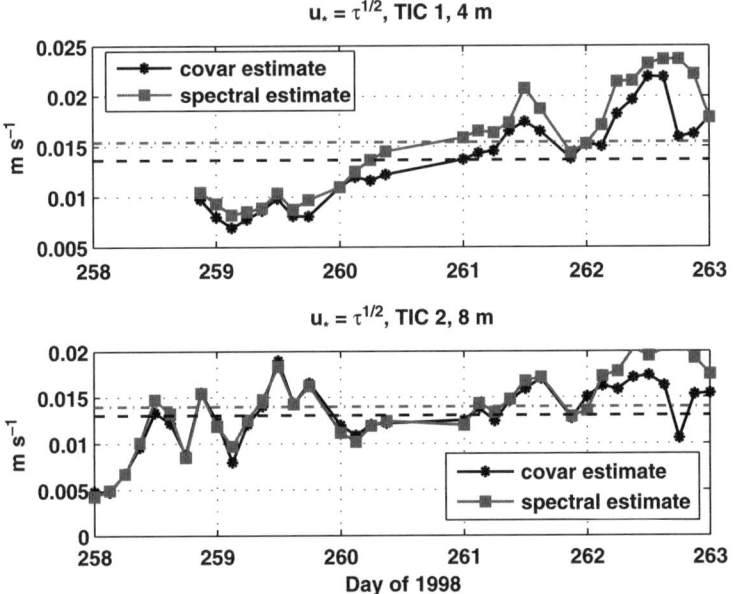

Fig. 3.10 Friction velocity as measured by direct covariance $\left(u_* = \left(\langle u'w'\rangle^2 + \langle v'w'\rangle^2\right)^{1/4}\right)$ averaged in 3-h blocks (black *) and derived from the w spectrum as described in the text (grey squares) at two levels near the end of the SHEBA project. Dashed (covariance) and dot-dashed (spectra) horizons show mean values (Adapted from McPhee 2004. With permission of the American Meteorological Society)

where $\phi = kS_w(k)$, evaluated at wave number $k = \gamma_* k_{max}$ where γ_* is a wave number in the $-2/3$ spectral range of the log-log w spectrum, normalized by the wave number at the maximum. We found the proportionality constant to be $c_\gamma = 0.48$. For SHEBA $\gamma_* = 2.5 (\log \gamma_* = 0.4)$ was consistently in the $-2/3$ range. Details of the derivation are presented in McPhee (2004). A comparison of friction velocity calculated via the spectral method (3.12) with direct covariance estimates for a time late in the SHEBA project is shown in Fig. 3.10.

3.7 Scalar Spectra

The analog between temperature variance conservation (3.6) and the TKE equation suggests that variance spectra of scalar variables might also provide useful insights into turbulent transfer processes in the IOBL. As with TKE, there is a fairly large body of observational evidence that temperature (heat) fluctuations are produced at scales corresponding to the energy-containing eddies, and dissipated at small scales. Consequently, there should exist an inertial subrange where the spectral density of temperature variance (S_T) depends only on three governing parameters: thermal dissipation rate, ε_T; wave number, k; and TKE characterized by ε. The governing

3.7 Scalar Spectra

parameters all have independent dimensions, so that we can again form a dimensionless group that will remain constant in the inertial subrange, and

$$S_T = \alpha_\Theta \varepsilon_T \varepsilon^{-1/3} k^{-5/3} \qquad (3.13)$$

where α_Θ is the thermal Kolmogorov constant. Edson et al. (1991) suggested a numerical value of 0.79 for α_Θ; we estimated a slightly higher value, 0.83, from statistics of nearly 400 3-h turbulent realizations at several levels in the IOBL during SHEBA (McPhee 2004).

In high-Reynolds number turbulent flows, it is often assumed that the eddy thermal diffusivity is nearly the same as eddy viscosity. If it is further assumed that thermal variance production and dissipation balance, then (3.6) provides a formula for the magnitude of vertical heat flux (divided by ρc_p):

$$\left| \langle w'T' \rangle \frac{\partial T}{\partial z} \right| = \frac{k_{\max} \langle w'T' \rangle^2}{c_\lambda u_*} = \varepsilon_T \qquad (3.14)$$

Given w and T spectra (and no other information) it is then possible to combine (3.14) with (3.13) and (3.12) for an estimate of the turbulent heat flux magnitude (although not direction) at a particular level. An example from near the end of the SHEBA experiment (adapted from McPhee 2004) comparing estimates made entirely from the spectra with direct covariance estimates is shown in Fig. 3.11.

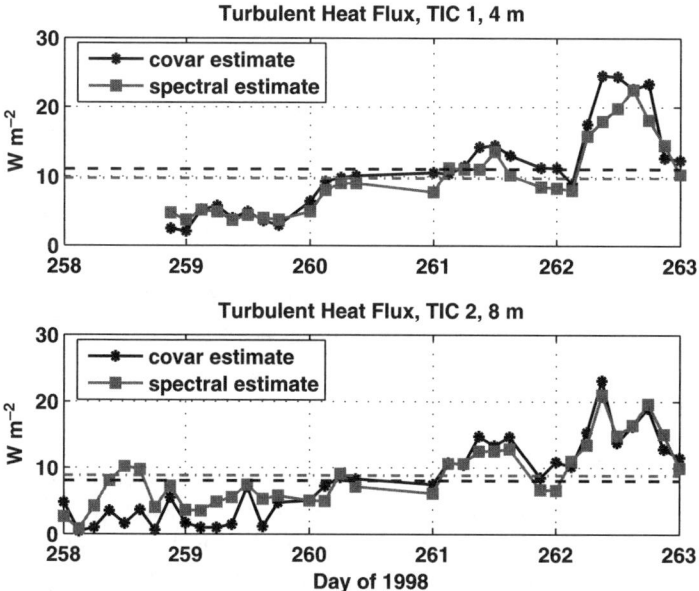

Fig. 3.11 Heat flux measured by direct covariance $(\rho c_p \langle w'T' \rangle)$ averaged in 3-h blocks (black $*$) and derived from the w and T spectra as described in the text (grey squares) at two levels near the end of the SHEBA project. Dashed (covariance) and dot-dashed (spectra) horizons show mean values (Adapted from McPhee 2004. With permission American Meteorological Society)

References

Barenblatt, G. I.: Scaling, Self-Similarity, and Intermediate Asymptotics. Cambridge University Press, Cambridge (1996)

Batchelor, G. K.: The Theory of Homogeneous Turbulence. Cambridge University Press, Cambridge (1967)

Bloomfield, P.: Fourier Analysis of Time Series: An Introduction. Wiley, New York (1976)

Bowker, A. H. and Lieberman, G. J.: Engineering Statistics. Prentice-Hall, Englewood Cliffs, NJ (1959)

Busch, N. E. and Panofsky, H. A.: Recent spectra of atmospheric turbulence. Quart. J. R. Met. Soc., 94, 132–147 (1968)

Edson, J. B., Fairall, C. W., Mestayer, P. G., and Larsen, S. E.: A study of the inertial-dissipation method for computing air-sea fluxes. J. Geophys. Res., 96, 10689–10711 (1991)

Efron, B. and Gong, G.: A leisurely look at the bootstrap, the jackknife, and cross-validation. Am. Statistician, 37, 36–48 (1983)

Emery, W. J. and Thomson, R. E.: Data Analysis Methods in Physical Oceanography, Second and Revised Edition. Elsevier, Amsterdam (2001)

Frisch, U.: Turbulence. Cambridge University Press, Cambridge (1995)

Hinze, J. O.: Turbulence, Second Edition. McGraw-Hill, New York (1975)

Jenkins, G. M. and Watts, D. G.: Spectral Analysis and Its Applications. Holden-Day, San Francisco (1968)

McPhee, M. G.: Greenland Sea Ice/Ocean Margin. EOS, Trans. Am. Geophys. Union, 64 (9), 82–83 (1983)

McPhee, M. G.: On the Turbulent Mixing Length in the Oceanic Boundary Layer. J. Phys. Oceanogr., 24, 2014–2031 (1994)

McPhee, M. G.: A spectral technique for estimating turbulent stress, scalar flux magnitude, and eddy viscosity in the ocean boundary layer under pack ice. J. Phys. Oceanogr., 34, 2180–2188 (2004)

McPhee, M. G. and Smith, J. D.: Measurements of the turbulent boundary layer under pack ice. J. Phys. Oceanogr., 6, 696–711 (1976)

McPhee, M. G. and Stanton, T. P.: Turbulence in the statically unstable oceanic boundary layer under Arctic leads. J. Geophys. Res., 101, 6409–6428 (1996)

McPhee, M. G., Maykut, G. A., and Morison, J. H.: Dynamics and thermodynamics of the ice/upper ocean system in the marginal ice zone of the Greenland Sea. J. Geophys. Res., 92, 7017–7031 (1987)

McPhee, M. G., Morison, J. H., and Nilsen, F.: Revisiting heat and salt exchange at the ice-ocean interface: Ocean flux and modeling considerations. J. Geophys. Res., doi:10.1029/2007JC004383, in press (2008)

Tennekes, H. and Lumley, J. L.: A First Course in Turbulence. MIT, Cambridge, MA (1972)

Nomenclature

c_p	Specific heat of seawater at constant pressure (approx. 4.0×10^3 J kg^{-1} K^{-1} for seawater at freezing)
$\langle w'T' \rangle$	Covariance estimate of kinematic turbulent heat flux
$\tau = \langle u'w' \rangle + i \langle v'w' \rangle$	Complex form of kinematic Reynolds stress
$u_* = \|\tau\|^{1/2}$	Friction speed (local)
Ω	Angular velocity of the earth (7.292×10^{-5} rad s^{-1})
δ_{ij}	Kronecker delta

Nomenclature

ε_{ijk}	Alternating tensor
ν	Molecular viscosity of cold seawater (approximately $1.8 \times 10^{-6}\,\text{m}^2\,\text{s}^{-1}$)
ν_T	Molecular thermal diffusivity of seawater (approximately $1.4 \times 10^{-7}\,\text{m}^2\,\text{s}^{-1}$)
TKE	Turbulent kinetic energy
P_S	Production of TKE by shearing
P_b	Production of TKE by buoyancy forces
D	Divergence of vertical flux of TKE plus w, p covariance
ε	TKE dissipation rate
$\langle w'b'\rangle = \frac{g}{\rho}\langle w'\rho'\rangle$	Buoyancy flux
λ	Scale of the energy-containing eddies (mixing length)
η	Kolmogorov length scale
υ	Kolmogorov velocity scale
ω	Frequency (s^{-1})
k	Angular wave number (rad m^{-1})
$S_{u,v,w}$	Variance (energy) spectral density for respective velocity components
α_ε	Kolmogorov constant for TKE dissipation (~ 0.5)
α_Θ	Kolmogorov constant for thermal dissipation (~ 0.8)
$k_{\max}$	Wave number at the maximum in the area-preserving w spectrum
c_λ	Proportionality constant in the relation $\lambda = c_\lambda/k_{\max}$ (0.85)
ϕ	$kS_w(k)$
γ_*	Nondimensional wave number in the inertial subrange (typically 2.5)
c_γ	Proportionality constant relating dimensionless spectrum to dimensionless wave number (~ 0.5) in the $-2/3$ log-log slope region

Chapter 4
Similarity for the Ice/Ocean Boundary Layer

Abstract: The concept of similarity is central to nearly all studies of fluid dynamics because it provides a means of reducing a whole class of flows to one set of equations, after nondimensionalizing with carefully chosen scales. By studying one instantiation of the class (say, in a laboratory or wind tunnel setting), results can be applied to other examples, perhaps less amenable to direct measurement. Familiar applications include testing of scale models to evaluate aerodynamic drag or lift. In this chapter, similarity in planetary boundary layers is examined in some detail. Relatively well known concepts (Monin-Obukhov similarity for buoyancy effects in the atmospheric surface layer and Rossby similarity for the drag exerted by the atmosphere on the surface) are described and used to illustrate the similarity between the atmospheric and oceanic boundary layers. We then combine these into a similarity theory for the IOBL stabilized by positive buoyancy flux at the surface (melting). The crucial parameters identified in the exercise, including important turbulence scales, then provide the rationale for development of the local turbulence closure model described in subsequent chapters.

4.1 The Surface Layer

From the first studies of wind structure near the surface in the atmosphere and in wind tunnels, there was strong evidence that the vertical profile of wind velocity was often nearly logarithmic, i.e., that

$$U \propto \log z$$

which implies that the wind shear is nearly inversely proportional to distance from the surface:

$$U_z = \frac{\partial U}{\partial z} \propto \frac{1}{z}$$

This was found to hold through the lower tens of meters in the atmosphere, where it was assumed (and verified by experiment) that the turbulent stress was nearly the same as the wind stress acting at the surface, so that friction speed is $u_0 \approx |\langle \mathbf{u'w'} \rangle|^{1/2}$, where the Reynolds stress is measured near the surface, typically at a standard level of 10 m.

If wind shear depends mainly on distance from the surface, on the local stress, and on viscosity of the fluid, then straightforward dimensional analysis (e.g., Barenblatt 1996) reveals that a dimensionless parameter including the dependent quantity wind shear and one or more of the independent governing parameters (z, u_{*0}, ν) will be a function of one other dimensionless group formed from the governing parameters since only two have independent dimensions. Consequently,

$$\frac{zU_z}{u_{*0}} = \Phi(\frac{zu_0}{\nu}) = \Phi(\mathrm{Re})$$

where Re is a Reynolds number formed with the friction velocity. Practically speaking, at the high Reynolds numbers typical of nearly all flows in the atmosphere or ocean, the Reynolds number dependence is minimal, and the *dimensionless wind shear* (for the neutrally stable) surface layer is

$$\phi_m = \frac{\kappa z U_z}{u_{*0}} = 1 \tag{4.1}$$

where $\kappa (= \Phi^{-1})$ is von Karman's constant, usually taken to be 0.4.

Note that in this development, there is no consideration of rotation, which manifests itself in the fluid equations in terms of the Coriolis parameter, f. If we accept that ν is unimportant, then we can reformulate the dimensional analysis above by substituting f for ν, where now the governing parameters have dimensions (following the notation of Barenblatt 1996): $[z] = L$, $[\mathbf{u}_{*0}] = LT^{-1}$, and $[f] = T^{-1}$. We again have a dimensionless group including the variable $\mathbf{U}_z$ that depends on one other dimensionless group (since there are three postulated governing parameters, two with independent dimensions):

$$\frac{z\mathbf{U}_z}{\mathbf{u}_{*0}} = \Phi(\frac{fz}{u_0}) = \Phi(\xi) \tag{4.2}$$

Note that the dimensionless group on the left is a vector (complex) relation indicating that shear and surface friction velocity need not be aligned. ξ is a dimensionless vertical coordinate, where the scale length is the planetary scale u_{*0}/f. A typical range for the planetary scale in polar oceans is 70–100 m. For dimensionless shear to be constant as a stipulation for remaining in the surface layer, ξ must be small since we observe angular shear relatively close to the interface. In the IOBL, the surface layer is much thinner than in the atmosphere. Assuming that air stress and water stress nearly balance in thin, freely drifting sea ice, and provided the *nondimensional* surface layer extent is about equal in both fluids (a basic tenet of PBL similarity discussed below), the ratio of dimensional surface layer extent will go as the square root of the density ratio, since $\rho_a u_{*a}^2 \approx \rho_w u_{*w}^2$. Thus the atmospheric surface layer extent will be roughly 30 times that of the ocean. At high latitudes, the ocean surface layer is often confined to within 2–4 m of the boundary, and, for example, the counterpart of the standard 10-m measurement height in the atmosphere would be only 30 cm from the interface in the ocean.

4.1.1 Mixing Length in the Neutral Surface Layer

As discussed in Chapter 3, one of the most useful simplifications for treating exchange of momentum in turbulent geophysical flows is the concept of eddy viscosity, which relates the turbulent momentum flux to mean current shear in the vertical

$$\langle w'u' \rangle = -KU_z$$

where K has units $[K] = L^2 T^{-1}$. If we assume that a reasonable scale velocity in the surface layer is u_{*0}, then it is immediately obvious from (4.1) that mixing length there is $\lambda_{sl} = \kappa z$ (where for the present z is taken to be the positive distance from the boundary).

4.1.2 The Law of the Wall and Surface Roughness Length

Integration of (4.1) leads to the logarithmic current (wind) profile in the neutrally stratified surface layer

$$U(z) = \frac{u_{*0}}{\kappa} \log z + \text{const} = \frac{u_{*0}}{\kappa} \log \frac{z}{z_0} \qquad (4.3)$$

where z_0 is in essence an integration constant, taken to be a length scale indicative of the roughness of the boundary. Equation (4.3) is often termed the "law of the wall" (LOW) and to the degree that it accurately describes the wind profile, measurements at two levels within the surface layer are sufficient to estimate the surface stress and roughness.

In laboratory flows, z_0 is found to be roughly 1/30th the size of the roughness elements on the surface. However, the undersurface of sea ice comprises a rather broad spectrum of roughness scales, with estimates ranging from several centimeters or more in old, highly deformed sea ice (e.g., the western Weddell Sea [McPhee 2008, in press]) and in the marginal ice zone where floes tend to break into smaller pieces with large edge area (McPhee et al. 1987), to sub-millimeter scales under young ice in the eastern Weddell (McPhee et al. 1999). Undeformed seasonal fast ice may be *hydraulically smooth* (e.g., Crawford et al. 1999) in which case, the undersurface roughness loses its dependence on the physical properties of the boundary, and instead depends only on friction velocity and molecular viscosity, i.e., $z_{0s} \cong (v/\hat{u}_0) e^{-2}$ (Hinze 1975).

In the surface layer, kinematic surface stress is often expressed as a function of wind speed squared:
$$\tau_0 = u_{*0}\boldsymbol{u}_{*0} = c_h U\boldsymbol{U}$$
where $\boldsymbol{U}$ is the velocity vector (relative to the surface) at distance h from the boundary, and c_h is the drag coefficient appropriate to level h. This is equivalent to specifying z_0 since
$$z_0 = he^{-\kappa/\sqrt{c_h}}$$
and it might seem unnecessary to consider undersurface roughness at all; indeed, most ice/ocean models specify quadratic drag between the ice and underlying ocean, sometimes with a constant turning angle (e.g., Hibler 1980). However, as explored in more detail below, quadratic stress is probably not appropriate except when applied to currents measured or modeled within the thin surface layer, confined to the upper few meters from the interface.

How best to specify undersurface roughness for large scale models is still largely unresolved, and because important ice/ocean exchange parameters depend on it, the problem continues to draw attention. From practical considerations, oceanographic instrumentation is most often deployed under reasonably smooth ice away from obvious obstacles, and a concern is that we thus systematically bias estimates of undersurface roughness (hence also stress and scalar flux estimates) downward. During the planning and analysis of SHEBA data, there was much emphasis on "scaling up" measurements made at one locality (the SHEBA station) to represent a regional area commensurate with the practical grid scale of a large scale numerical model. This is particularly germane in the case of surface roughness and will be explored in more detail below.

4.1.3 Monin-Obukhov Similarity

The LOW (4.3) is valid only when buoyancy effects are negligible. In the atmospheric surface layer this precludes a large percentage of possible states; for example, strong diurnal heating from incoming shortwave radiation, or nocturnal surface cooling from outgoing longwave radiation. In the former case, buoyant air parcels formed near the surface tend to enhance turbulence and increase the efficiency of exchange, thus reducing shear; whereas in the latter case, a temperature inversion may form near the surface, much reducing the frictional coupling and increasing near surface shear. The ice/ocean interface exhibits analogous conditions when there is melting or freezing (although rarely to the same intensity as found in the daily cycle of the mid-latitude atmosphere).

The impact of buoyancy on the surface layer wind shear was first investigated systematically by Obukhov (1971, English translation; see also Businger and Yaglom 1971). For the purposes here, dimensional analysis suffices to define the main features of the problem. We postulate that wind (current shear) in the surface layer depends on friction velocity, scaled distance from the interface, and buoyancy flux:
$$U_z = F(u_{*0}, \kappa z, \langle w'b'\rangle_0)$$

4.1 The Surface Layer

Again using the formalism introduced in Section 3.5, we have

$$[U_z] = T^{-1} \qquad \text{dependent quantity}$$

$$[\kappa z] = L$$
$$[u_{*0}] = LT^{-1} \qquad \text{governing parameters}$$
$$[\langle w'b'\rangle_0] = L^2 T^{-3}$$

Three governing parameters, two with independent dimensions, imply that a dimensionless group including wind shear will be an as yet undetermined function of one other dimensionless group, formed from powers of the governing parameters. The parameter exponents are determined by solving

$$[\langle w'b'\rangle_0] = [\kappa z]^{p_1}[u_{*0}]^{p_2}$$
$$L{:}2 = p_1 + p_2$$
$$T{:}-3 = p_2$$

whence

$$\left[\frac{\kappa z \langle w'b'\rangle_0}{u_{*0}^3}\right] = 0$$

so that the counterpart of (4.1) when buoyancy flux is important is

$$\phi_m = \frac{\kappa z U_z}{u_{*0}} = \phi_m\left(\frac{\kappa z \langle w'b'\rangle_0}{u_{*0}^3}\right) = \phi_m\left(\frac{z}{L_0}\right) = \phi_m(\zeta) \qquad (4.4)$$

where L_0 is the Obukhov length based on surface (interface) momentum and buoyancy flux. If L_0 is positive (assuming z is positive displacement from the boundary), turbulence is suppressed by buoyancy and *vice versa* if L_0 is negative.

The form of $\phi_m(\zeta)$ has been studied extensively for the atmospheric surface layer. For example, Businger et al. (1971) suggested the following empirical formula covering a broad range of atmospheric conditions observed from a 32-m tower over flat terrain in Kansas[1]:

$$\phi_m(\zeta) = \begin{cases} 1 + 4.7\zeta & L_0 \geq 0 \\ (1 - 15\zeta)^{-1/4} & L_0 < 0 \end{cases} \qquad (4.5)$$

Lettau (1979) studied near-surface profiles in the stably stratified atmosphere over the Antarctic ice cap and suggested that the dimensionless shear had some curvature, i.e.

$$\phi_m(\zeta) = (1 + 5\zeta)^{3/4} \quad L_0 \geq 0 \qquad (4.6)$$

---

[1] Note that data from even this relatively tall tower in the atmosphere would correspond to measurements made within about the first meter of the IOBL.

In general, this approach to interpreting fluxes of momentum (and additionally scalar variables) in terms of mean quantity profiles in the surface layer has become known as Monin-Obukhov similarity.

As long as rotation is unimportant, (4.4) provides a formula for mixing length in the surface layer

$$\lambda_{sl} = \frac{\kappa z}{\phi_m}$$

which combined with an empirical formula like (4.5) shows how the vertical extent of the energy containing eddies depends on buoyancy. If ζ is large and positive (i.e., $L_0 > 0$; $L_0 < z$), mixing length is severely reduced from its neutral values. Conversely, it increases dramatically if ζ is large and negative (statically unstable). This observation introduces an important principle for understanding the impact of various scales on turbulence: *it is the smaller scale that governs*.

4.2 The Outer Layer

As distance from the boundary increases, rotation in the IOBL can no longer be ignored. Figure 4.1, showing horizontal currents at several levels averaged for 5 h during a storm at the AIDJEX Pilot Study station in 1972 (McPhee and Smith 1976), is representative of the IOBL velocity structure often observed under sea ice. The dashed curve connects velocity measurements at various levels from 2 to 32 m from the ice. Below 32 m, angular shear was small and velocity at that level measured relative to the drifting ice was close to the apparent bottom velocity measured by an acoustic bottom reference system. Thus absolute current at 32 m was small, and vectors drawn from that depth represent the actual velocity in the IOBL. In this representation, the x-axis is along the direction of the 2-m current, which is taken to be the direction of stress acting on the interface (opposite the surface stress acting on the IOBL). The dashed vector indicates where the near surface current is aligned 45° *cum sole* from the surface stress, taken to be approximately the upper limit of the Ekman layer. Figure 4.1b presents the same information in terms of vertical profiles of the velocity. Despite large current shear in the upper few meters, the integral of velocity (i.e., volume transport) is nearly all in the positive y direction, at right angles to the surface stress.

In the short term, velocity measurements at a particular site with (relative) currents emanating from a particular upstream direction are often influenced by underice morphology (e.g., note the irregularities between 4 and 16 m in Fig. 4.1). This is especially evident in multiyear pack ice including different ice types and thicknesses. With a long enough record, however, we can examine flows from many directions with different stress conditions. During the ISPOL experiment we monitored currents in the upper ocean more or less continuously with an acoustic Doppler profiler (ADP) for most of the month of December. We considered every 3-h average data set, sampled the profiles at 2-m intervals from 10 to 30 m, and nondimensionalized the complex (2-d) velocity vectors by dividing by the complex current at 30 m. Results, shown in Fig. 4.2, show a reasonably well developed Ekman spiral in

4.2 The Outer Layer

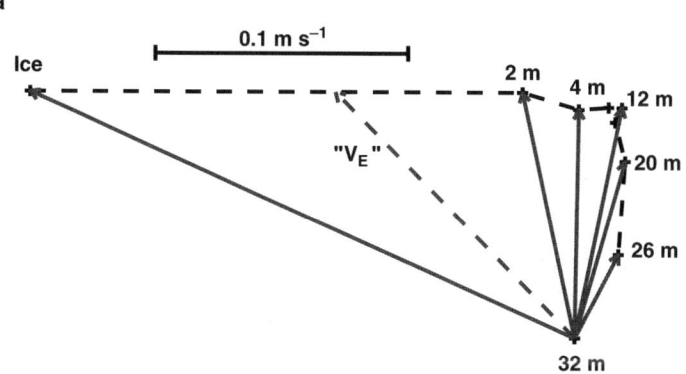

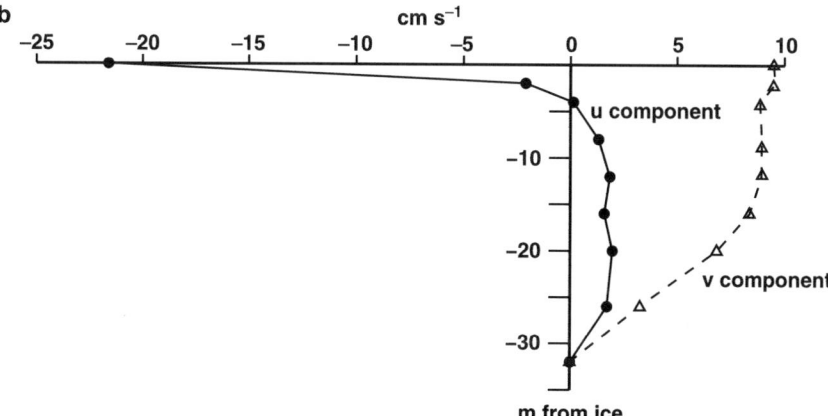

Fig. 4.1 Currents averaged for 5 h during a storm at the 1972 AIDJEX Pilot Study in the western Arctic. **a.** Plan view hodograph. The dashed vector represents velocity where the direction between stress and velocity is $\pi/4$. **b.** Corresponding profiles in the upper 32 m of the water column, relative to the current measured at 32 m

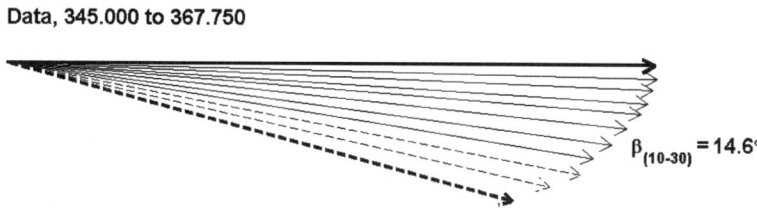

Fig. 4.2 Average nondimensional current hodograph (plan view) of complex currents measured relative to the ice and sampled every 2 m from 10 to 28 m, after nondimensionalizing by the 30-m current (McPhee 2008, in press)

currents beyond the surface layer. In the southern hemisphere, current backs clockwise with increasing depth to an observer on the ice.

4.2.1 Similarity for Turbulent Stress in the Outer Layer

As suggested by dimensional analysis in Section 4.1, dimensionless vector current shear likely depends on z scaled by the planetary length u_{*0}/f as indicated by (4.2). We used this similarity to infer that scales in the PBLs of the atmosphere and ocean differed by a factor of about 30. Here we extend the earlier analysis to similarity for the outer boundary layer, including consideration of turbulent stress.

Start with the steady, horizontally homogeneous OBL equation

$$if u = \frac{\partial \tau}{\partial z} \qquad (4.7)$$

and choose nondimensional variables for stress and vertical coordinate

$$T = \frac{\tau}{u_{*0} u_{*0}}$$

$$\xi = \frac{z}{H}$$

where the obvious choice for scaling stress is the applied surface stress. Substituting into (4.7), we have the nondimensional momentum equation

$$iU = \frac{\partial T}{\partial \zeta} \qquad (4.8)$$

provided

$$U = \frac{fH}{u_{*0}} \cdot \frac{u}{u_{*0}} \qquad (4.9)$$

If H is the planetary scale (u_{*0}/f), then obviously the dimensionless velocity is u/u_{*0}, but here we will retain the generality implicit in (4.9).

For first-order closure

$$\tau = K \frac{\partial u}{\partial z}$$

$$T = K_* \frac{\partial U}{\partial \xi}$$

where the nondimensional eddy viscosity is $K_* = \frac{K}{fH^2}$. If we hypothesize, following Ekman, that the z dependence of eddy viscosity is negligible in the outer layer, then differentiation of (4.8) leads to a second order ordinary differential equation for T:

4.2 The Outer Layer

$$\frac{i}{K_*}T = \frac{\partial^2 T}{\partial \xi^2} \quad (4.10)$$

with boundary conditions for the IOBL (with z positive upward)

$$T(0) = 1$$
$$T(-\infty) = 0$$

and solution simply

$$T = e^{\delta \xi} \quad (4.11)$$

where

$$\delta = (\pm i/K_*)^{1/2} \quad (4.12)$$

where the sign preceding i depends on the hemisphere (+northern). The exponential argument is complex, meaning that it both rotates and attenuates stress as depth increases (Fig. 4.3); i.e., the Ekman spiral pertains to stress as well as velocity.

The problem is formulated here in terms of stress rather than velocity since almost by definition, there is little variation in stress within the relatively thin surface layer, and we can surmise that near the boundary the Reynolds stress spiral is insensitive to variation in eddy viscosity. This contrasts with the strong shear apparent in the velocity profile. Significant turning in the Reynolds traction vector with increasing depth has been observed repeatedly during ice station experiments. An example is illustrated by measurements (Fig. 4.4) from the Ice Station Weddell experiment (McPhee and Martinson 1994). During a storm that lasted about two days, strong southerly winds blew the station northward, setting up a well developed and relatively steady IOBL. We calculated zero-lag covariance of vertical with horizontal velocity components for each 15-min flow realization, then averaged these over the course of the storm to get mean kinematic stress estimates at 5 levels ranging from 4 to 24 m below the ice/water interface. Results clearly show substantial leftward deflection and attenuation of the Reynolds stress with increasing depth.

The spiral profile shown in Fig. 4.4 is from

$$\tau(z) = u_{*0}\boldsymbol{u}_{*0} \exp\left(\frac{1}{\sqrt{2K_*}} (1-i) |f| z/u_{*0} \right) \quad (4.13)$$

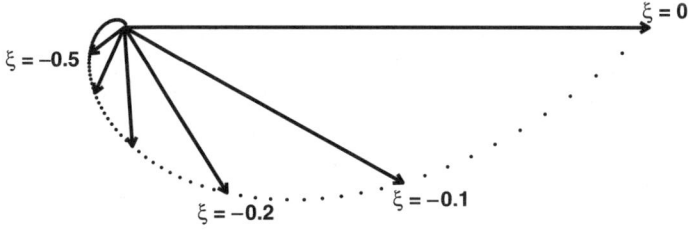

Fig. 4.3 The nondimensional stress profile illustrating the Ekman stress spiral

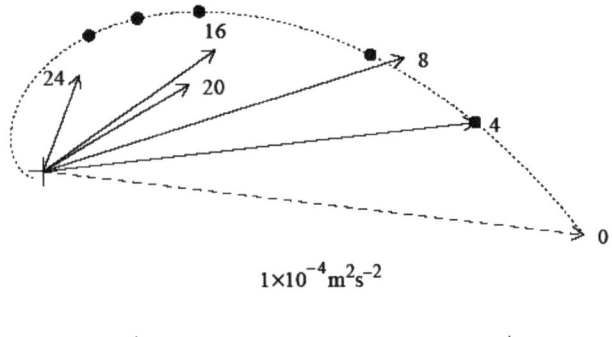

Fig. 4.4 Turbulent stress measured as five levels during a storm at Ice Station Weddell in 1992. The dotted curve is the simple similarity model for stress, with dimensional parameters chosen so that the stress at 4 m matches observed. The dashed vector labeled "0" indicates the inferred boundary stress (From McPhee and Martinson 1994. With permission American Association for the Advancement of Science)

where $K_* = 0.02$ (based primarily on estimates from AIDJEX measurements, see McPhee 1981) and the surface stress is chosen in order to make

$$\tau(z = -4 \text{ m}) = \overline{\langle u'w' \rangle_4} + i\overline{\langle v'w' \rangle_4} \tag{4.14}$$

Buoyancy flux was negligible, so $\xi = fz/u_{*0}$. In this case the friction velocity magnitude is $u_{*0} = 0.012 \text{ m s}^{-1}$ and consequently the estimate of constant eddy viscosity is $K_{sim} = K_* u_{*0}^2 / f = 0.021 \text{ m}^2 \text{ s}^{-1}$.

Suppose that in the neutral IOBL, mixing length increases linearly from the surface until it reaches a limiting value $\lambda_{max} = \kappa z_{sl}$ and the product of friction speed at that level with λ_{max} is the maximum eddy viscosity. The dimensionless maximum mixing length $\Lambda_* = f\lambda_{max}/u_{*0}$ will also constitute a similarity parameter. If we further assume that attenuation of friction velocity in the surface layer is negligible, so that $K_{max} = u_{*0}\lambda_{max}$, then $K_* = \Lambda_*$.[2]

4.2.2 Rossby Similarity for the Neutral IOBL

The conceptual model implicit in the Ekman solution is that for a given stress condition, mixing length and eddy viscosity do not vary much through the outer part of the IOBL. Yet it is obvious that from the perspective of velocity shear (as opposed to kinematic) stress, linear dependence of λ in the surface layer is quite important, and the total shear will depend strongly on surface roughness, z_0. Consider Ekman velocity and ice velocity (relative to the bottom of the boundary layer) depicted in

---

[2] In fact there is appreciable stress attenuation in the surface layer and Λ_* is somewhat larger than K_* as discussed later.

4.2 The Outer Layer

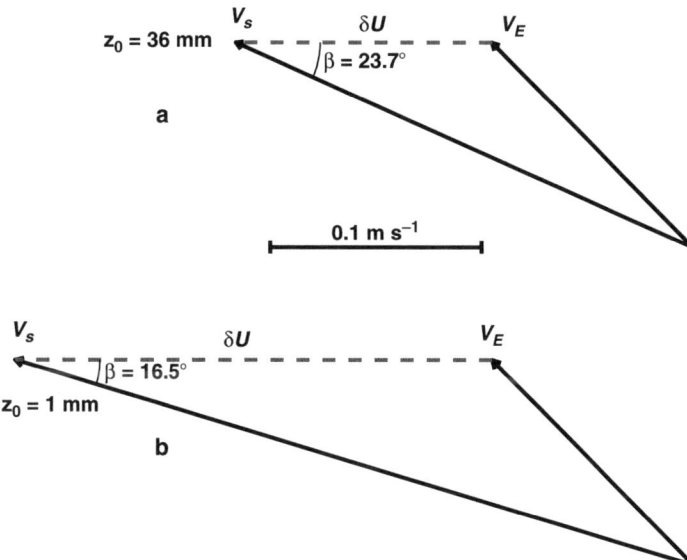

Fig. 4.5 Impact of different undersurface roughness lengths on surface velocity for the same surface stress and Ekman layer velocity. Vectors labeled V_E are velocity at the top of the Ekman layer as described in the text (Adapted from Fig. 4.1)

Fig. 4.1. We estimated (McPhee and Smith 1976) that u_{*0} was close to 0.01 m s^{-1} when the entire IOBL was considered, but that the near surface velocity profile was distorted by local underice topography. However, since the ice moved as a unit over the entire area observable from the ice station, the total shear across the IOBL does reflect an integration over a much larger region than the relatively smooth area surrounding the instrument mast. It is thus reasonable to assume that the difference between the vector labeled u_E and u_{ice} (δU in Fig. 4.5a) is representative of shear across the regional surface layer, from which it is straightforward to calculate z_0 assuming that the surface layer extends to the point where $\kappa|z| = u_{*0}\Lambda_*/f$. For the conditions shown, z_0 is about 0.04 m, which is fairly typical of multiyear pack ice. In the first year ice of the Weddell Sea, z_0 is closer to 1 mm (McPhee et al. 1999), so all else being equal, δU is larger as shown in Fig. 4.5b. The ice moves about 40% faster for the same stress, with a significant reduction in turning angle.

If u_{*0} varies instead of z_0, then clearly u_E and δU will change. However, in the similarity sense, the nondimensional Ekman velocity remains the same. What changes is the scaled value for the surface roughness $\xi_0 = f z_0 / u_{*0}$, so that the impact of increased stress is to decrease the scaled surface roughness, and increase the nondimensional surface layer shear. In other words, for fixed undersurface roughness, the impact of increased stress is to lower the effective drag coefficient magnitude and to decrease the angle of turning between surface stress and velocity.

Relating boundary stress to surface (ice) velocity is analogous to relating wind stress at the surface to the geostrophic wind aloft. For the atmosphere, a method called *Rossby similarity* has been used to quantify the effects described qualitatively above. Let $V_s = F(u_{*0}, z_0, f)$ be the surface velocity relative to the ocean beyond the IOBL. Again by dimensional analysis

$$\frac{V_s}{u_{*0}} = U_0\left(\frac{u_{*0}}{fz_0}\right) \tag{4.15}$$

where the dimensionless grouping of governing parameters is the ratio of the planetary scale to the boundary roughness scale, which in the atmospheric literature is often called the surface friction Rossby number, Ro_* (e.g., Blackadar and Tennekes 1968).

We can exploit the simple conceptual model of nondimensional stress as a complex exponential, combined with a surface layer in which eddy viscosity varies linearly with ξ, to investigate the functional form of (4.15). The integral of (4.11) from $-\infty$ to the base of the surface layer, $\xi_{sl} = -\Lambda_*/\kappa$, provides the nondimensional velocity at the top of the Ekman layer:

$$U_E = -i\delta e^{\delta \xi_{sl}} \tag{4.16}$$

In the surface layer (small $|\xi|$) a Taylor-series expansion for the exponential provides

$$T \simeq 1 + \delta\xi$$

and

$$U_E \simeq -i\delta(1 + \delta\xi_{sl})$$

and

$$\frac{\partial U}{\partial \xi} = \frac{1}{-\kappa\xi}(1 + \delta\xi)$$

from which integration to the boundary ($\xi = \xi_0 = -fz_0/u_{*0}$) provides the nondimensional surface velocity

$$U_0 = U_E + \frac{1}{\kappa}\left[\ln\frac{\xi_{sl}}{\xi_0} + \delta(\xi_{sl} - \xi_0)\right] \tag{4.17}$$

with $|\xi_0| \ll |\xi_{sl}|$ the real and imaginary components of U_0 are

$$\text{Re}(U_0) = \frac{1}{\kappa}\left[\log\frac{u_{*0}}{fz_0} + \log\frac{\Lambda_*}{\kappa} + \frac{\kappa}{\sqrt{2\Lambda_*}} - 1 - \sqrt{\frac{\Lambda_*}{2\kappa^2}}\right] = \frac{1}{\kappa}[\log Ro_* - A]$$

$$\text{Im}(U_0) = \frac{1}{\kappa}\left[-\frac{\kappa}{\sqrt{2\Lambda_*}} - \sqrt{\frac{\Lambda_*}{2\kappa^2}}\right] = \frac{-1}{\kappa}B \tag{4.18}$$

These complicated looking expressions thus reduce to a formula for the "geostrophic" drag law for sea ice

4.2 The Outer Layer

$$\frac{V_s}{u_{*0}} = U_0 = \frac{1}{\kappa}(\log Ro_* - A \mp iB) \quad (4.19)$$

where the sign of the imaginary term depends on the hemisphere (negative for northern). This relates the total change in (vector) velocity across the IOBL to the vector friction velocity at the boundary, provided the well mixed layer is not shallow compared with the depth to which turbulence is active, and boundary buoyancy flux is small (L_0 large).

The imaginary part of (4.19) is constant and provides guidance for estimating the value of the similarity parameter Λ_* from

$$B = \left(\frac{\kappa}{\sqrt{2\Lambda_*}} + \sqrt{\frac{\Lambda_*}{2\kappa^2}}\right) \quad (4.20)$$

From results of a wind/free-drift analysis of AIDJEX data, we estimated (McPhee 1981) that $B \approx 2.1$ which implies $\Lambda_* = 0.024$. The corresponding value for A is then about 2.3. It should be noted that atmospheric estimates of B obtained by comparing geostrophic wind aloft with surface friction velocity are often considerably higher. But these analyses depend on friction velocity measured near typically smooth boundaries, thus ignoring the impact of larger roughness features that might affect geostrophic drag, and also do not consider the impact of relatively low inversion heights. Still, the value derived from the AIDJEX analysis should be considered only approximate. We have found from analyzing and modeling IOBL data from several different sources that a reasonable value is slightly larger: $\Lambda_* = 0.028$, but acknowledge that this could easily depart from the "true" constant by as much as 10–15%.

In terms of a customary quadratic drag relationship, $|\tau_0| = c_g V_s^2$ the drag coefficient is $c_g = U_0^{-2}$ but according to (4.19), drag does not follow a strictly quadratic power relation—in fact, the exponent in the relation $\tau \propto V_s^n$ was close to $n = 1.7$ rather than 2, for the AIDJEX stations in free drift (McPhee 1979). For a typical sea-ice undersurface roughness of 0.03 m, with $A = 2.3$ and $B = 2.1$, the drag coefficient magnitude and turning angle for ice speeds ranging from 0.05 to 0.5 m s^{-1} are plotted in Fig. 4.6.

4.2.3 Similarity for the Stably Stratified IOBL

When there is stabilizing buoyancy flux from ice melting, the vertical scales of turbulence are reduced by gravitational force, and mixing is confined to shallower depths than for neutral stability. This has two important effects. First, fresher water will begin to form a new pycnocline within the well mixed layer that existed when melting began. This may inhibit subsequent mixing below its newly established level, and we find that in the central Arctic, seasonal pycnoclines generally form by mid-to-late July as meltwater mixes into the upper ocean. Second, by confining momentum transfer to a shallower layer, it will reduce effective drag on the ice

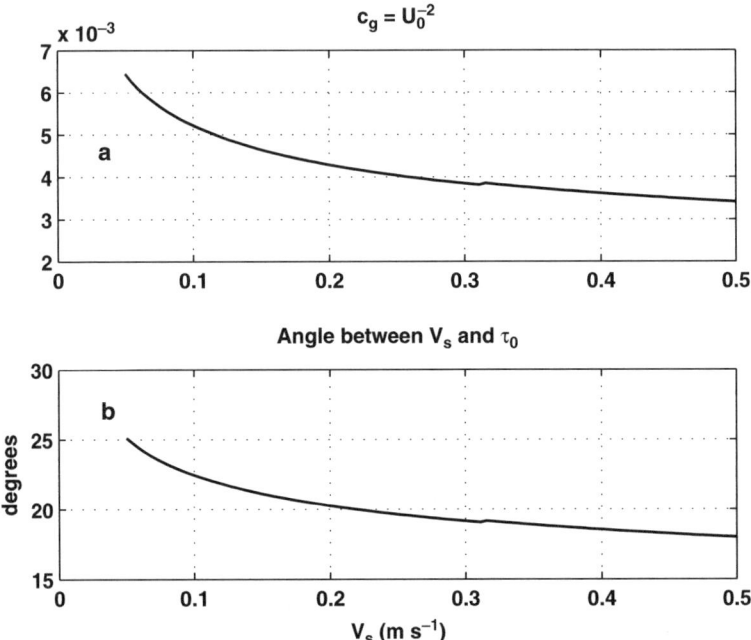

Fig. 4.6 Drag coefficient from Rossby similarity. **a** magnitude. **b** Turning angle between stress and surface velocity

underside, and possibly modify the amount of Ekman deflection in ice drift. In this section, we consider an extension to the similarity approach that includes buoyancy flux (McPhee 1981, 1983). We seek a similarity solution to the Ekman stress equation that includes the effects of surface buoyancy flux.

We postulated earlier that there is a maximum mixing length in the neutral IOBL given by $\lambda_{max} = \Lambda_* u_{*0}/f$ where Λ_* is a "universal" similarity parameter. As stabilizing buoyancy affects turbulence, we anticipate that turbulence scales including λ_{max} will decrease. Its lower limit follows from considering the simplified TKE equation with the surface layer approximation of constant flux

$$P_s + P_b = \tau \cdot \frac{\partial u}{\partial z} - \langle w'b' \rangle_0 = \frac{u_{*0}^3}{\lambda} - \langle w'b' \rangle_0 = \varepsilon \quad (4.21)$$

Dividing through by P_s

$$1 + \frac{P_b}{P_s} = 1 - \frac{\lambda}{\kappa L} = 1 - R_f = \frac{\varepsilon \lambda}{u_{*0}^3} \quad (4.22)$$

The key here is that the ratio of buoyancy flux to shear production, called the *flux Richardson number* has a limiting critical value, R_c, beyond which turbulence ceases to exist in laboratory flows. A commonly accepted value is around 0.2. Thus for turbulence to be viable

4.2 The Outer Layer

$$R_f = \frac{\langle w'b'\rangle \lambda}{u_{*0}^3} < R_c \approx 0.2 \quad \Rightarrow \quad \lambda_{max} < R_c \kappa L \tag{4.23}$$

and we have established limits for λ_{max}:

$$\lambda_{max} \to \Lambda_* u_{*0}/f \quad \text{for} \quad L \to \infty \tag{4.24}$$
$$\lambda_{max} \to R_c \kappa L \quad \text{for} \quad L \to 0^+$$

A simple expression with these asymptotes is half the harmonic mean of the limits in (4.24):

$$\lambda_{max} = \eta_*^2 \Lambda_* u_{*0}/f \tag{4.25}$$

where

$$\eta_* = \left(1 + \frac{\Lambda_* u_{*0}}{\kappa R_c f L}\right)^{-1/2} = \left(1 + \frac{\Lambda_* \mu_*}{\kappa R_c}\right)^{-1/2} \tag{4.26}$$

μ_* is a stability parameter that represents the ratio of the planetary length scale to the Obukhov length. It is now possible to re-evaluate the master length scale H, since

$$K_* = \frac{K}{fH^2} = \frac{u_{*0}^2 \eta_*^2}{f^2 H^2} \Lambda_* = \Lambda_* \tag{4.27}$$

where the last equality follows from the stipulation that all flows within the class being considered (neutral and stably stratified PBLs) are similar. Thus the similarity scales are

$$\begin{aligned}
\text{Length:} &\quad \eta_* u_{*0}/f \\
\text{Velocity:} &\quad u_{*0}/\eta_* \\
\text{Eddy viscosity:} &\quad (u_{*0}\eta_*)^2/f \\
\text{Kinematic stress:} &\quad u_{*0} u_{*0}
\end{aligned} \tag{4.28}$$

where both velocity and kinematic stress scales are vector (complex) quantities.

For the IOBL stabilized by positive buoyancy flux at the boundary, we anticipate that $V_s = F(u_{*0}, z_0, f, L_0)$ where L_0 is the Obukhov length based on boundary fluxes, and that the nondimensional relation will be of the form

$$U_0 = \frac{V_s}{u_{*0}} = U_0\left(\frac{u_{*0}}{fz_0}, \frac{u_{*0}}{fL_0}\right) = U_0(Ro_*, \mu_*) \tag{4.29}$$

A typical average value of friction speed for perennial sea ice in the Arctic is about 7 mm s^{-1}. For a specified basal melt rate, the boundary buoyancy flux may be calculated following the formulas developed in Chapter 6. It turns out that for typical sea-ice parameters, with $u_{*0} = 7$ mm s^{-1}, the magnitude of μ_* is about the same as melt rate expressed in centimeters per day (Fig. 4.7). In Fig. 4.8, the dimensional stress is plotted for the same surface stress, $\tau = u_{*0}^2 = 4.9 \times 10^{-5}$ m^2 s^{-2}, but three different values of μ_*: 0 (neutral), 5, and 25. The dimensionless stress is, of course, the same in all cases (Fig. 4.3) because the scaled boundary layers are

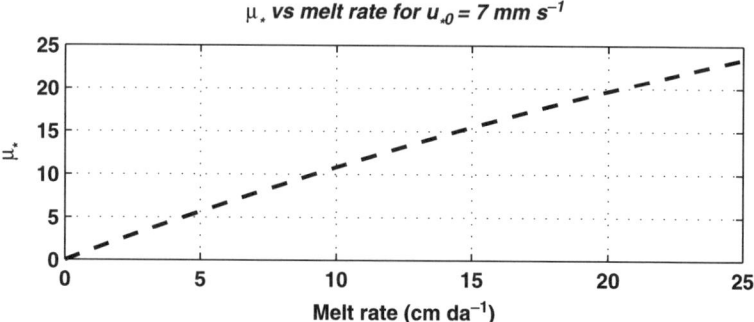

Fig. 4.7 The ratio of planetary scale to Obukhov length (μ_*) for a range of basal melt rates and fixed interface stress typical of average values for Arctic sea ice

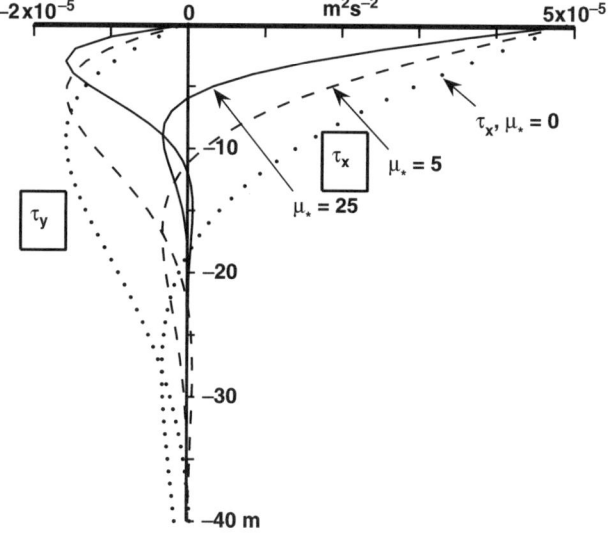

Fig. 4.8 Profiles of kinematic stress components for three values of the stability parameter, μ_*

similar, but even though u_{*0}/f remains fixed, the magnitude of dimensional z varies as η_*, which is 1, 0.60, and 0.32, respectively, for the given range of μ_*. Note that for the maximum μ_*, equivalent roughly to about 25 cm da^{-1} melt rate, the stress is confined to the upper 15 m or so of the water column, which would therefore limit the depth to which mixing of the meltwater would reach. A new pycnocline would rapidly form at about that level. Even with a relatively modest melt rate of about 5 cm da^{-1}, the depth of mixing would be significantly reduced.

Velocity in the surface layer is obtained in the same way as before by integrating dimensionless shear from the velocity at the top of the outer layer

$$\frac{V_s}{u_{*0}} = U_0/\eta_* \tag{4.30}$$

using the Taylor-series expansion of the exponential stress

$$\frac{\partial U}{\partial \xi} = \frac{T}{K_*} \cong \frac{-\eta_*}{\kappa \xi}(1 - \beta \mu_* \eta_* \xi)(1 + \delta \xi) \qquad \xi > \Lambda_*/\kappa \qquad (4.31)$$

So for the entire IOBL, the similarity equations are

$$T = \exp(\delta \xi) \qquad (4.32)$$

and

$$U(\xi) = \begin{cases} -i\delta \exp(\delta \xi) & \xi \leq \xi_{sl} \\ U_E - \frac{\eta_*}{\kappa}\left[\log \frac{\xi}{\xi_{sl}} + (\delta - a)(\xi - \xi_{sl}) - \frac{a\delta}{2}\left(\xi^2 - \xi_{sl}^2\right)\right] & \xi > \xi_{sl} \end{cases} \qquad (4.33)$$

where $a = \beta \mu_* \eta_* = \frac{\kappa(1-\eta_*)}{\eta_* \Lambda_*}$, $\delta = (\pm i/\Lambda_*)^{1/2}$, and $\xi_{sl} = -\Lambda_*/\kappa$.

Equations (4.32) and (4.33) constitute the similarity theory with parameters $R_c = 0.2$ and $\Lambda_* = 0.028$.

The dimensionless surface velocity is

$$U_0 \simeq -i\delta(1 + \delta \xi_{sl}) + \frac{\eta_*}{\kappa}\left[\log \frac{\xi_{sl}}{\xi_0} - (a - \delta)\xi_{sl} - \frac{a\delta}{2}\xi_{sl}^2\right] \qquad (4.34)$$

where $\xi_0 = -\frac{fz_0}{\eta_* u_{*0}}$. The geostrophic drag relation is

$$\frac{V_s}{u_{*0}} = \frac{U_0}{\eta_*} \qquad (4.35)$$

which as before may be manipulated into a Rossby-similarity form, except that now the parameters A and B depend on μ_*

$$\frac{V_s}{u_{*0}} = \frac{1}{\kappa}[\log Ro_* - A(\mu_*) - iB(\mu_*)] \qquad (4.36)$$

4.3 IOBL Similarity and the Atmospheric Boundary Layer

4.3.1 Dimensionless Shear

Velocity in the surface layer is related to stress by eddy viscosity that depends on both distance from the boundary and the Obukhov length

$$K = \frac{\kappa |z| u_{*0}}{1 + \beta |z|/L_0} \qquad (4.37)$$

with nondimensional form

$$K_* = \frac{-\kappa\xi}{\eta_*}(1-\beta\mu_*\eta_*\xi)^{-1} \qquad \xi > \Lambda_*/\kappa \qquad (4.38)$$

For similarity, K must match the outer-layer eddy viscosity at $z_{sl} = -\frac{\eta_* u_{*0}\Lambda_*}{f\kappa}$. From this it follows that β, considered an empirical factor in the discussion of the atmospheric surface layer, is here an internal parameter in the problem that depends weakly on the stability parameter μ_*

$$\beta = \left(\frac{1}{R_c} + \frac{\kappa}{\mu_*\Lambda_*}\right)(1-\eta_*) \qquad (4.39)$$

We can examine the behavior of β in the surface layers as follows. The dimensionless shear is

$$\phi(\zeta) = \frac{\kappa|z|}{u_{*0}}\frac{\partial u}{\partial z} = \frac{-\kappa\xi}{\eta_*}\frac{\partial U}{\partial \xi} = \frac{T}{K_*} = T(1-\beta\eta_*\mu_*\xi) \qquad (4.40)$$

and the identity $\zeta = -\eta_*\mu_*\xi$ along with the Taylor-series approximation for T provide

$$\phi(\zeta) = (1+\beta\zeta)\left(1 - \frac{\delta\zeta}{\eta_*\mu_*}\right) \qquad (4.41)$$

Since β is a function of μ_* in the similarity theory, so is ϕ, which is plotted for two values ($\mu_* = 50, 100$) in Fig. 4.9. Also shown are the empirical formulas for $L_0 > 0$ from (4.5) and (4.6). For $\mu_* = 50$, the similarity estimate is virtually indistinguishable from Lettau's (1979) empirically fitted function.

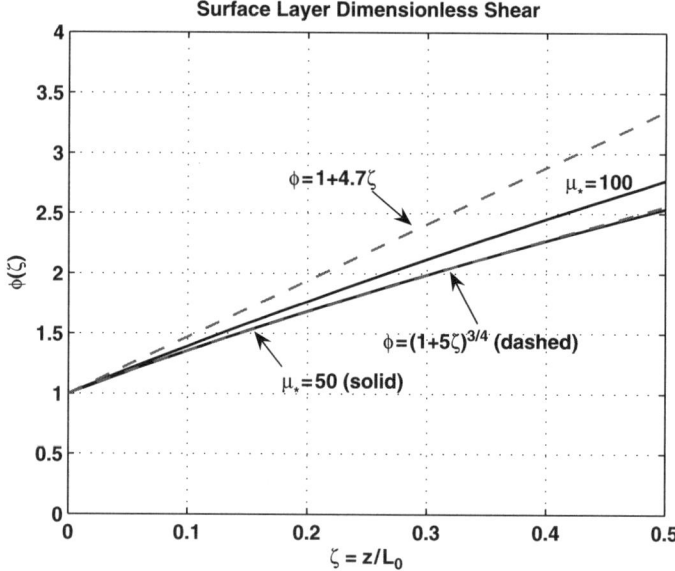

Fig. 4.9 Surface-layer dimensionless shear equation as a function of z/L_0. Solid curves are from the IOBL similarity theory for two different values of μ_*. Dashed curves are from Businger et al. (1971) (upper) and from Lettau (1979) (lower)

4.4 Ice-Edge Bands

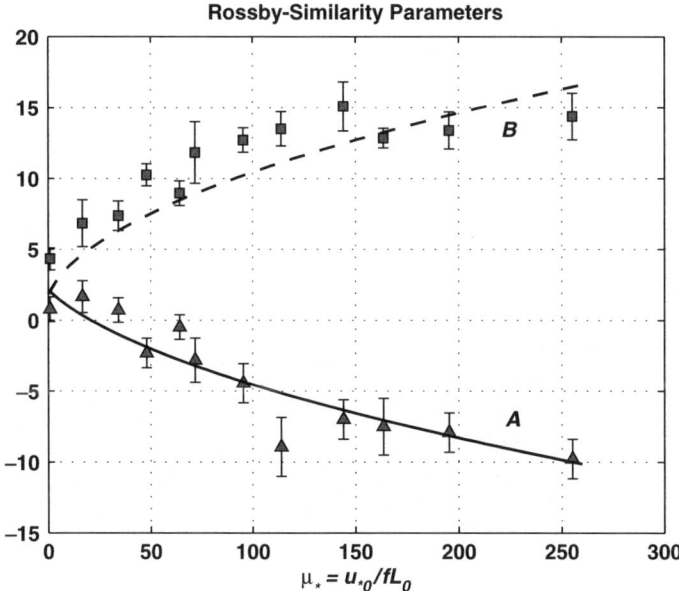

Fig. 4.10 Comparison of similarity estimates of the buoyancy dependent stability parameters with empirical estimates from Clarke and Hess (1974) using geostrophic wind and surface fluxes measured over level ground in Australia

4.3.2 The Rossby-Similarity Parameters for Stable Stratification

Clarke and Hess (1974) investigated empirically the relationship between geostrophic wind and surface flux conditions over flat terrain in Australia (the Wangara data). Their results are plotted in Fig. 4.10 as a function of the ratio of the planetary scale to the Obukhov length (their definition differs by a factor of κ), along with the functions derived from (4.36).

4.4 Ice-Edge Bands

The reader may wonder at the importance of showing how the similarity theory for the underice boundary layer satisfies some of the atmospheric constraints discussed above. After all, as shown in Fig. 4.7, it would take unrealistically large melt rates to approach the levels of μ_* seen routinely in the atmosphere. This is true for statically unstable conditions as well. In general, the diurnal buoyancy cycle observed in the mid-latitude atmosphere is much more intense in its impact on the boundary layer than in the ocean. For pack ice away from marginal ice zones or large polynyas,

Fig. 4.11 Two views of an ice-edge band observed during the 1983 MIZEX project in the Greenland Sea from the bridge of the R/V *Polarbjörn*

neutral scaling ($\eta_* = 1$) will most likely suffice most of the time.[3] Nevertheless, the basic scaling ideas behind the relatively simple similarity theory described in this chapter, particularly the buoyancy scaling factor, η_*, form the basis for more complex modeling of the IOBL presented later. And the basic point behind similarity

---

[3] When the well mixed layer is shallow compared to the dynamic IOBL depth (nominally $0.4\eta_* u_{*0}/f$) interaction with fluid in the pycnocline will modify the drag relation.

is that the IOBL and atmospheric boundary layer have much in common, thus the extensive observational base available for the atmosphere is applicable to the ocean.

Similarity scaling of IOBL drag may help explain the formation of ice-edge bands (McPhee 1983; Mellor et al. 1986). There is often a fairly abrupt front observed in ocean temperature near the edge of the ice pack in marginal ice zones like that found in the Greenland or Bering Seas. When off-ice winds drive the main pack across such a front, melting is rapid, and it is commonly observed that relatively thin bands of ice drift away (downwind) from the main pack. An example observed during the 1983 MIZEX project in the Greenland Sea marginal ice zone is shown in Fig. 4.11. We encountered this band after sailing south from the main pack for some distance in open water. While no environmental data are available from this site, it is reasonable to assume that the southward drifting ice was in water well above freezing. Suppose for example that $u_{*0} \sim 0.01 \, \text{m s}^{-1}$ and $\mu_* \sim 20$ (corresponding roughly to a melt rate of 40 cm per day). Then application of (4.34) and (4.35) implies that the leading edge of the ice pack would travel about 8–10 km per day farther than similar ice with no melting. In this view, the edge band forms because water is cooled rapidly behind the advancing ice, so slower melting means less stratification and consequently ice following the leading edge moves slower. The width of the band probably reflects the size of the internal boundary layer that would form from the leading edge—which is far beyond the scope of the similarity theory.

Alternative mechanisms for ice-edge band formation have been suggested, notably wave radiation pressure on the trailing edge (Martin et al. 1983; Wadhams 1983). The photograph in Fig. 4.11 seems to indicate, however, that the following edge feathers out, which would be more consistent with the "slippery water" effect than wave radiation pressure.

References

Barenblatt, G. I.: Scaling, Self-Similarity, and Intermediate Asymptotics. Cambridge University Press, Cambridge (1996)

Blackadar, A. K. and Tennekes, H.: Asymptotic similarity in neutral planetary boundary layers. J. Atmos. Sci., 25, 1015–1019 (1968)

Businger, J. A. and Yaglom, A. M.: Introduction to Obukhov's paper on 'Turbulence in an atmosphere with a non-uniform temperature'. Boundary-Layer Meteorol., 2, 3–6 (1971)

Businger, J. A., Wyngaard, J. C., Izumi, Y., and Bradley, E. F.: Flux-profile relationships in the atmospheric surface layer. J. Atmos. Sci., 28, 181–189 (1971)

Clarke, R. H. and Hess, G. D.: Geostrophic departure and the functions A and B of Rossby-Number similarity theory. Boundary-Layer Meteorol., 7, 267–287 (1974)

Crawford, G., Padman, L., and McPhee, M.: Turbulent mixing in barrow strait. Continental Shelf Res., 19, 205–245 (1999)

Hibler, W. D. III.: A dynamic thermodynamic sea ice model. J. Phys. Oceanogr., 10, 815–846 (1980)

Hinze, J. O.: Turbulence, Second Edition. McGraw-Hill, New York (1975)

Lettau, H. H.: Wind and temperature profile prediction for diabatic surface layers including strong inversion cases. Boundary-Layer Meteorol., 17, 443–464 (1979)

Martin, S., Kauffman, P., and Parkinson, C.: The movement and decay of ice edge bands in the winter bering sea. J. Geophys. Res., 88, 2803–2812 (1983)

McPhee, M. G.: The effect of the oceanic boundary layer on the mean drift of sea ice: Application of a simple model. J. Phys. Oceanogr., 9, 388–400 (1979)

McPhee, M. G.: An analytic similarity theory for the planetary boundary layer stabilized by surface buoyancy. Boundary-Layer Meterol., 21, 325–339 (1981)

McPhee, M. G.: Turbulent heat and momentum transfer in the oceanic boundary layer under melting pack ice. J. Geophys. Res., 88, 2827–2835 (1983)

McPhee, M. G.: Physics of early summer ice/ocean exchanges in the western Weddell Sea during ISPOL, Deep-Sea Res., II, doi:10.1016/j.dsr2.2007.12.022, in press (2008)

McPhee, M. G., Kottmeier, C., and Morison, J. H.: Ocean heat flux in the central Weddell Sea in winter. J. Phys. Oceanogr., 29, 1166–1179 (1999)

McPhee, M. G. and Martinson, D. G.: Turbulent mixing under drifting pack ice in the Weddell Sea. Science, 263, 218–221 (1994)

McPhee, M. G. and Smith, J. D.: Measurements of the turbulent boundary layer under pack ice. J. Phys. Oceanogr., 6, 696–711 (1976)

McPhee, M. G., Maykut, G. A., and Morison, J. H.: Dynamics and thermodynamics of the ice/upper ocean system in the marginal ice zone of the Greenland Sea. J. Geophys. Res., 92, 7017–7031 (1987)

Mellor, G. L., McPhee, M. G., and Steele, M.: Ice-seawater turbulent boundary layer interaction with melting or freezing. J. Phys. Oceanogr., 16, 1829–1846 (1986)

Obukhov, A. M.: Turbulence in an atmosphere with a non-uniform temperature. Boundary-Layer Meteorol., 2, 7–29 (1971)

Wadhams, P.: A mechanism for the formation of ice edge bands. J. Geophys. Res., 88, 2813–2818 (1983)

Nomenclature

ϕ_m	Dimensionless current (wind) shear
κ	Von Kármán's constant (0.4)
ξ	Dimensionless vertical coordinate (z/H)
Φ	Complex dimensionless current shear
z_0	Surface roughness length
L_0	Obukhov length scale based on boundary fluxes $L_0 = u_{*0}^3/(\kappa \langle w'b' \rangle_0)$
ζ	Surface layer dimensionless vertical coordinate $\zeta = z/L_0$
λ_{sl}	Surface-layer mixing length, $\kappa z/\phi_m$
λ_{max}	Maximum mixing length in the IOBL
U_0	Complex dimensionless surface velocity $\mathbf{V}_s/\mathbf{u}_{*0}$
Ro_*	Surface-friction Rossby number $(u_{*0}/f)/z_0$
A and B	Rossby-similarity functions
Λ_*	Similarity parameter
R_c	Critical flux Richardson number (0.2)
μ_*	Stability parameter $(u_{*0}/f)/L_0$
η_*	Similarity scale factor for the stably stratified IOBL

Chapter 5
Turbulence Scales for the Ice/Ocean Boundary Layer

Abstract: Understanding the scales of turbulence in the IOBL is the central issue in developing reasonable models for transfer of properties between the ice cover and the underlying ocean. This chapter presents several examples from field observations that shed light on the impact of both stress and buoyancy on turbulence in the IOBL, and use them to develop a heuristic approach to specifying the mixing length. A key in this development is the apparent connection between the inverse wave number at the peak of the area-preserving w spectrum and the master length scale for turbulence. As discussed in Section 3.6, the concept was first explored for the IOBL using data from the 1972 AIDJEX Pilot Study (McPhee and Smith 1976). Figure 5.1, adapted from that work, shows our estimates of eddy viscosity based on admittedly crude analysis of the spectra observed during an AIDJEX storm, analyzed in the manner suggested by Busch and Panofsky (1968), and compared with calculations from one of the first attempts at large eddy simulation for the atmospheric boundary layer (Deardorff 1972). This was far from conclusive; however, later measurements tended to confirm that basic approach.

5.1 Neutral OBL Scales

Several ice-station experiments have provided data on spectral characteristics of turbulence at various levels in the IOBL. Here we concentrate on two experiments that illustrate the connection between the inverse of the wave number at the peak of the area preserving vertical velocity variance spectrum and λ, as introduced in Section 3.6.

5.1.1 Ice Station Weddell

The hodograph of Reynolds stress traction vectors depicted in Fig. 4.4 was obtained during a storm at Ice Station Weddell in 1992 (McPhee and Martinson 1994). For

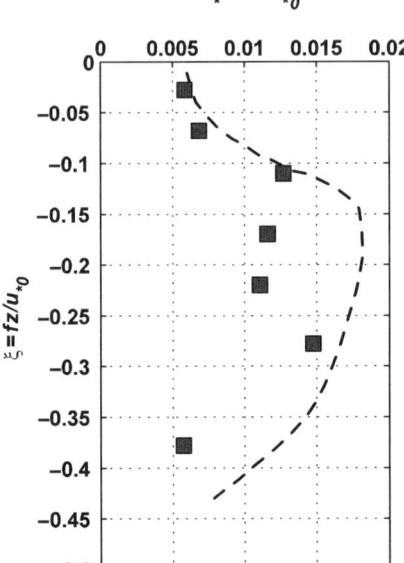

Fig. 5.1 Estimates of dimensionless eddy viscosity at several levels under drifting pack ice during the AIDJEX 1972 Pilot Study, compared with an early large-eddy-simulation (Deardorff 1972) (Adapted from McPhee and Smith 1976)

spectral calculations (McPhee 1994), data were segregated into 1-h time series, from which the average velocity was used to rotate the vector components into a streamline coordinate system in which mean vertical ($\overline{w}$) and cross-stream ($\overline{v}$) components vanish, and $U = \overline{u}$. After linearly detrending u, spectra were calculated following the procedure described in Section 3.5. Spectral components were then bin averaged in evenly spaced bins of $\log_{10}(k)$ where k is the angular wave number. The 1-h time series were overlapped by half for better statistics at higher wave numbers. Twelve estimates for each of 5 TICs ranging in depth from 4 to 24 m (TIC 3 at 12 m malfunctioned), were then further averaged on a common $\log_{10}(k)$ grid for a total time span of 6 h. Resulting spectra for TIC 5 at 20 m have been discussed in Section 3.5 and are shown Fig. 3.9. We postulated that the presence of a reasonably well defined region where the log-log slope of the area-preserving spectra was $-2/3$, accompanied by a region in which the ratio $S_{ww} \approx 4/3 S_{uu}$, indicated isotropy at small scales in the inertial subrange of the flow.

To test the hypothesis that $\lambda = c_\lambda / k_{\max}$, we estimated the peak in each average w spectrum by determining the maximum in a high-order polynomial fitted to the spectral density estimates. Following Busch and Panofsky (1968), the spectrum at 4 m was fitted to a function of the form

$$\frac{kS_{ww}(k)}{u_*^2} = \frac{A(k/k_{\max})}{1 + 1.5(k/k_{\max})^{5/3}} \tag{5.1}$$

5.1 Neutral OBL Scales

We found that A was close to unity. By assuming that $\lambda = \kappa |z|$ at 4 m (the uppermost turbulence cluster), the estimate for c_λ was about 0.85, slightly larger than the neutral values suggested by Busch and Panofsky (0.8). From this we constructed an estimate of mixing length based on the wave number at the peak in the average w spectra at 5 levels during the ISW storm (McPhee and Martinson 1994):

$$\lambda_{\text{peak}} = c_\lambda / k_{\max} \tag{5.2}$$

An independent estimate of the mixing length may be made from turbulence spectra by assuming that locally TKE production rate equals dissipation, where ε is derived from spectral levels in the inertial subrange (Hinze 1975)

$$\varepsilon^{2/3} = \frac{3}{4\alpha_\varepsilon} S_{ww}(k) k^{5/3} \tag{5.3}$$

where $\alpha_\varepsilon = 0.51$ is the Kolmogorov constant for turbulent kinetic energy. The process is essentially an inversion of the inertial-dissipation method for estimating stress in the atmospheric surface layer (e.g., Edson et al. 1991), where instead of using ε and κz (i.e., λ in the surface layer) to estimate stress, we use u_* obtained from the covariance measurements and ε from the inertial subrange to calculate λ:

$$\lambda_\varepsilon = u_*^3 / \varepsilon = \left[\frac{3k_\varepsilon S_{ww}(k_\varepsilon)}{4\alpha_\varepsilon}\right]^{-3/2} \frac{u_*^3}{k_\varepsilon} \tag{5.4}$$

where u_* is the local friction velocity, k_ε is a wave number chosen in the $-2/3$ region of the w spectrum, and $S_{ww}(k_\varepsilon)$ is the spectral density evaluated at k_ε.

With measurements of temperature-velocity covariance and TKE dissipation, an analogous *thermal mixing length*, λ_T may be derived by equating temperature variance production with thermal dissipation (see Section 3.7)

$$\left|\langle w'T'\rangle \frac{\partial T}{\partial z}\right| = \frac{\langle w'T'\rangle^2}{u_* \lambda_T} = \varepsilon_T = \frac{\varepsilon^{1/3}}{\alpha_T} S_{TT}(k_\varepsilon) k_\varepsilon^{5/3} \tag{5.5}$$

where $\alpha_T = 0.79$ (Edson et al. 1991) is the Komogorov constant for temperature variance, and $S_{TT}(k_\varepsilon)$ is the spectral level for temperature variance at a specific wave number in the inertial subrange.

Estimates of the three different mixing lengths for the ISW-92 storm, summarized along with TKE and thermal variance dissipation rates in Fig. 5.2, are generally similar throughout the IOBL, and all depart markedly from surface layer ($\kappa|z|$) scaling. These results suggest that mixing length remains relatively constant through the IOBL, once past a relatively shallow surface layer where it is proportional to distance from the boundary. This observation forms the basis for the hypothesis that eddy viscosity in the well mixed portion of the IOBL beyond the surface layer may be represented by

$$K = \lambda u_* \tag{5.6}$$

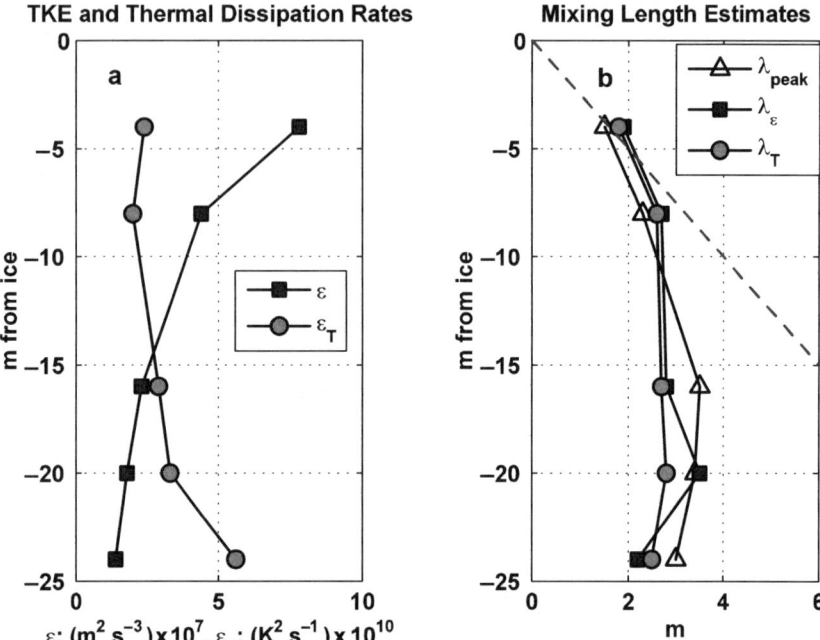

Fig. 5.2 a TKE and thermal variance dissipation rates versus depth at ISW 92. **b** Estimates of mixing length derived from spectral peaks (λ_{peak}), from the TKE balance assuming production equals dissipation (λ_ε), and from the thermal variance conservation equation (λ_T) (Adapted from McPhee and Martinson 1994. With permission American Association for the Advancement of Science)

where u_* is the *local* friction velocity. This differs from the similarity model discussed in Chapter 4 in that K is allowed to vary in the outer layer, while we assume that λ remains relatively constant.

Without resorting to similarity scaling or mixing-length arguments, it is possible to estimate a representative eddy viscosity during the ISW storm directly from Ekman theory. If the kinematic stress magnitude is exponential, i.e., $\tau = \tau_0 e^{az}$, we can estimate the parameters τ_0 and a by the linear regression of $\log \tau$ versus z. Results of the fitting are shown in Fig. 5.3. From the Ekman solution

$$K_{\text{fit}} = f/(2a^2)$$

where a is the slope of the semilogarithmic fit. With $a = 0.051$, $K_{\text{fit}} = 0.026 \, \text{m}^2 \, \text{s}^{-1}$.

The rather unique data set from the ISW storm also provides a credible estimate of the scalar thermal eddy diffusivity averaged through the entire IOBL, by relating directly the vertical averaged kinematic heat flux ($\overline{\langle w'T' \rangle}$) and thermal gradient. With an average heat flux of less than $10 \, \text{W m}^{-2}$ during the ISW storm, a rough estimate of the expected thermal gradient in the well mixed layer may be made by assuming that eddy scalar diffusivity is comparable to eddy viscosity ($0.02 \, \text{m}^2 \, \text{s}^{-1}$):

5.1 Neutral OBL Scales

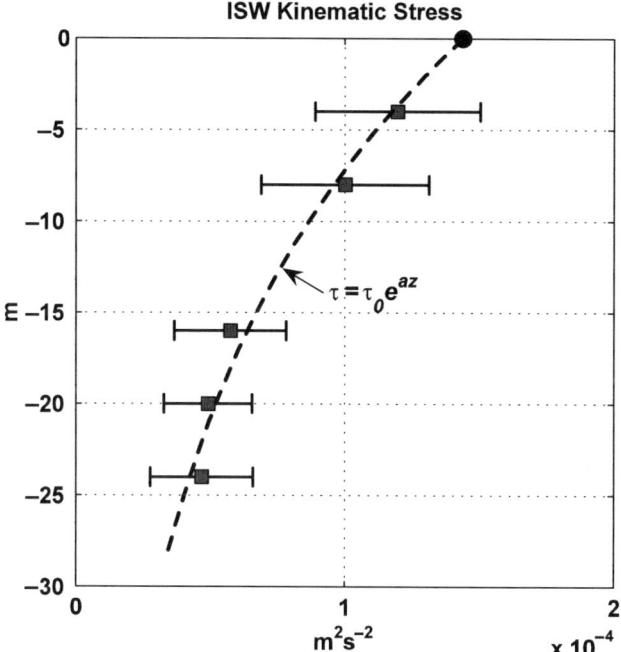

Fig. 5.3 Kinematic stress at five levels during the ISW storm. The error bars represent +1 std. deviation of the covariance estimates of kinematic Reynolds stress in each 15-min "realization." The dashed line is a log-linear fit of the stress magnitudes as described in the text (Adapted from McPhee and Martinson 1994. With permission American Association for the Advancement of Science)

$$\frac{\partial T}{\partial z} \sim \frac{\overline{-\langle w'T'\rangle}}{K} \sim 10^{-4}\,\mathrm{K\,m^{-1}}$$

which, of course, is very small. The stated accuracy of the Sea-Bird oceanographic thermometers used during ISW was $\pm\,0.01\,°\mathrm{C}$, consistent with the sort of variability we observed in IOBL temperature using the factory calibrations. However, by identifying times when measured heat flux was near zero, we were able to apply constant corrections to each thermometer that eliminated any potential temperature gradient during that time. With these corrections it was then possible to estimate time series of temperature gradient in the IOBL by linear regression of the adjusted temperatures. The vertically averaged heat flux and negative thermal gradient (Fig. 5.4) are strongly correlated, and provide an estimate of the average thermal diffusivity, which is $K_T = 0.018\,\mathrm{m^2\,s^{-1}}$.

The various estimates of discrete and bulk eddy viscosity/diffusivities are summarized in Fig. 5.5. They are generally consistent, and support the hypothesis that a mixing length derived from the inverse of the wave number at the peak in the w spectrum may be a powerful tool for understanding turbulence scales in the IOBL.

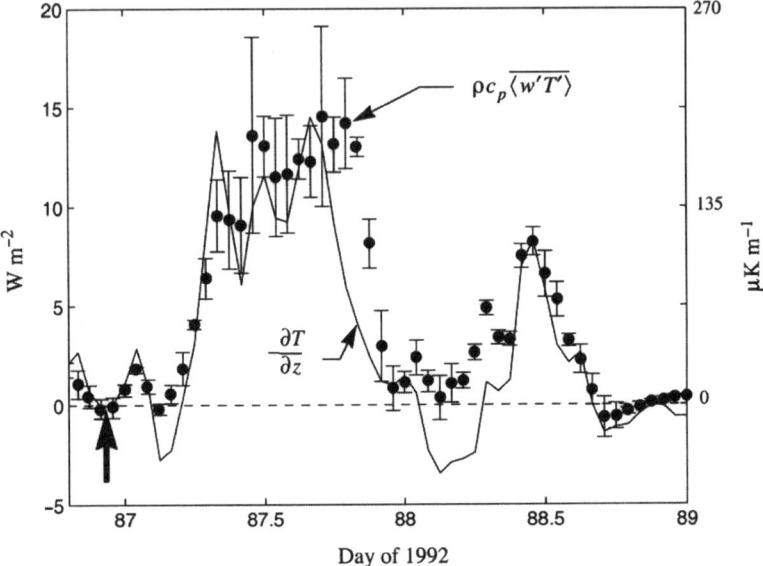

Fig. 5.4 Comparison of average measured heat flux at five levels in the IOBL at ISW (solid circles with error bars representing ± one standard deviation) with the negative of the thermal gradient. The heavy arrow indicates the time chosen to calculate temperature calibration adjustments so that the potential temperature gradient was zero for no heat flux (From McPhee and Martinson 1994. With permission American Association for the Advancement of Science)

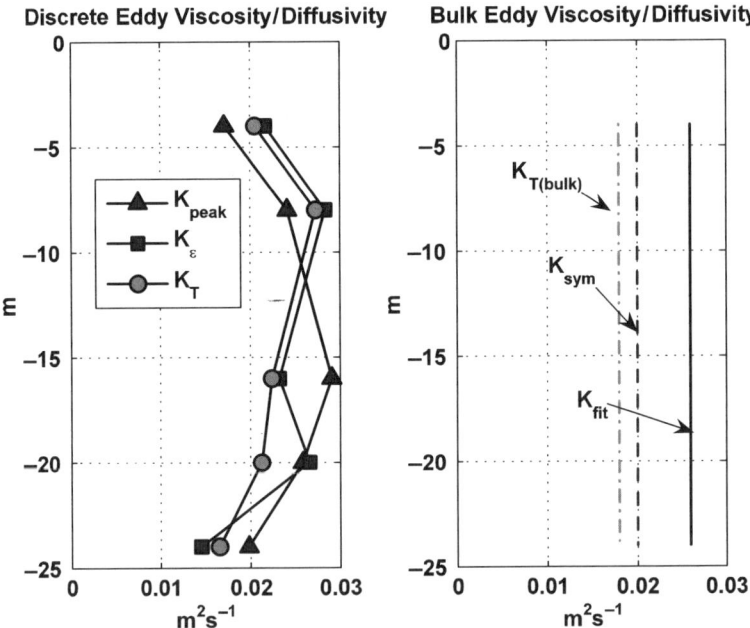

Fig. 5.5 Various estimates of eddy viscosity and eddy thermal diffusivity made at discrete depths via spectral techniques, compared with bulk values as described in the text. Discrete values were obtained by multiplying various λ estimates (From Fig. 5.2b) by the local u_*

5.1.2 Ice Station Polarstern

As discussed previously, during the 2004–2005 ISPOL drift in the western Weddell Sea, we deployed turbulence clusters equipped with ADVOcean current meters in the upper ocean under a heterogeneous multiyear ice floe. The ADVs measure velocity by gauging the Doppler shift of scattered sound in a small (approximately 1×2 cm) cone separated from the sensors by about 18 cm. As long as there are adequate sound scatterers in the fluid, the current meters have virtually no lower threshold velocity with little interference from the apparatus. When flow energy is low, this makes them far superior to the earlier "Smith-rotor" mechanical current meter triplets for measuring current characteristics.

Over the course of the month-long turbulence-mast deployment, several configurations were utilized, with two or three TICs at depths ranging from 1 to 10 m from the ice-water interface (McPhee 2008, in press). As with previous analyses, the TIC data streams were divided into 15-min realizations, rotated into a streamline reference frame with covariance statistics and variance spectra calculated for each realization. Smoothed spectral estimates were averaged in evenly spaced $\log(k)$ bins, then further averaged in 3-h blocks. A typical average, area-preserving w spectrum is shown in Fig. 5.6, with a high-order polynomial fit from which the maximum is determined, as indicated by the dot-dashed line.

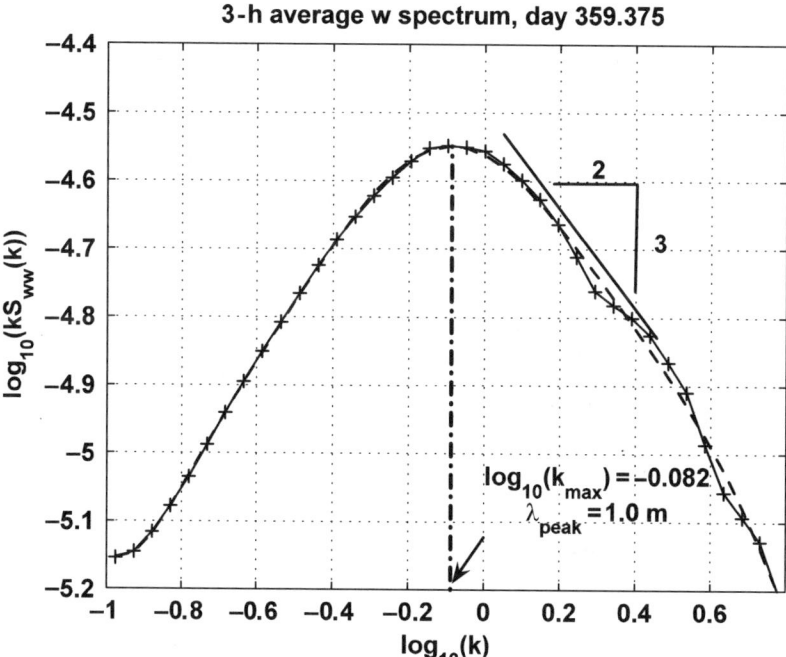

Fig. 5.6 Three-hour average spectrum from ISPOL centered at 0900 UT on 23 December 2004. The dashed line is a high-order polynomial fit, from which the maximum is determined where the derivative is zero (From McPhee 2008, in press)

Since we were able to measure Reynolds stress with the ADVs to much lower levels than possible previously, we reasoned that the ISPOL turbulence data set might provide guidance in determining just how far the z-dependence of mixing length extended in a similarity sense. With neutral stratification the maximum mixing length should be the smaller of $\Lambda_* u_*/|f|$ or $\kappa |z|$, depending on whether the measurement level is in the surface layer or outer layer (where we are making the tacit assumption that the surface layer is defined as the region where mixing length varies with distance from the boundary). At 67°S, $|f| = 1.33 \times 10^{-4}\,\text{s}^{-1}$, and with our estimate of $\Lambda_* = 0.028$, for measurements made 4 m from the boundary, λ would be the smaller of $\sim 200 \times u_*$ or 1.6 m, where the tilde acknowledges that u_* may differ somewhat from u_{*0}. Ideally then, if we plotted λ versus u_* it should increase linearly until it reached $u_* \approx 8\,\text{mm}\,\text{s}^{-1}$ and then remain about constant for higher values. In other words, for friction velocity values less than $8\,\text{mm}\,\text{s}^{-1}$, 4 m is beyond the surface layer and the planetary scale dominates, whereas for greater values, the geometric scale $\kappa |z|$ rules.

For all of the 3-h turbulence samples during the first ISPOL deployment (1–25 December), we assembled a scatter plot of $\lambda = 0.85/k_{\max}$ versus $u_* = |\langle u'w'\rangle + i\langle v'w'\rangle|^{1/2}$ (Fig. 5.7). Despite fairly large sample-to-sample variation, the regression

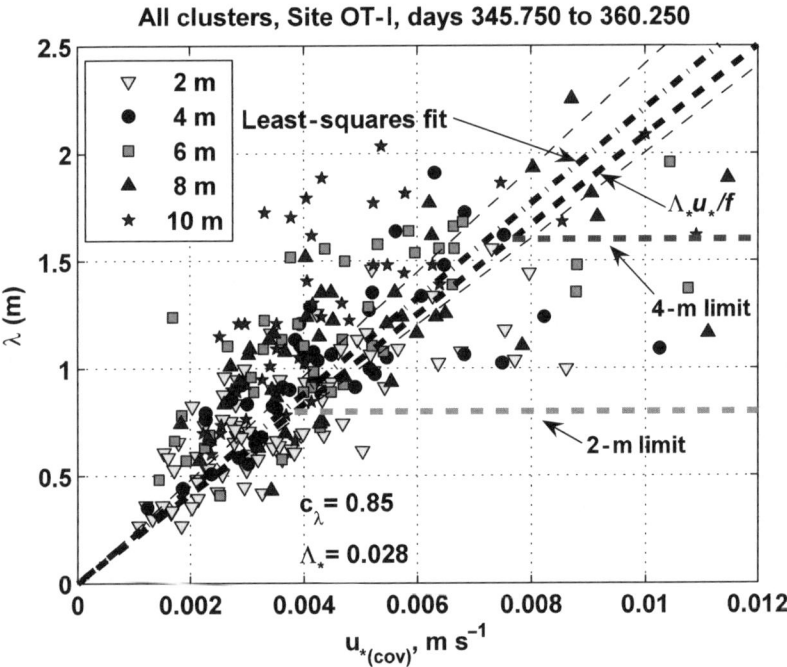

Fig. 5.7 Scatter plot of mixing length λ versus u_* for all 3-h averages prior to the ISPOL Christmas-Day breakup. The heavy dot-dashed line is a least-squares linear fit through the origin with 95% confidence interval indicated by the light dashed lines. The heavy dashed line indicates the dynamic (planetary) maximum mixing length. The grey dashed lines indicate the "geometric" limits, $\kappa |z|$, at 2 and 4 m, respectively (From McPhee 2008, in press) (see also Plate 16 in the Colour Plate Section)

of λ against u_* provides convincing evidence that for low values of stress, it is often the planetary rather than the geometric scale that governs turbulence, even relatively close to the boundary.

5.2 The IOBL with Stabilizing Boundary Buoyancy Flux

As discussed in Section 4.2.3, in compact pack ice it is rare to encounter sustained conditions where boundary buoyancy flux has major impact on ice/ocean drag characteristics or near surface turbulence scales.[1] This is particularly true for manned research stations, which are typically sited with an eye for survivability. The 1984 Marginal Ice Zone Experiment (MIZEX) north of Fram Strait in the Greenland Sea was an exception. For this ship-supported drift, we hoped to encounter a range of conditions elucidating the behavior of pack ice as it encounters the open ocean. After drifting with a multiyear floe for about three weeks in late June and early July with ocean conditions fairly representative of the Arctic summer ice pack, northerly winds blew the station southward across a front that marked the edge of an eddy previously visible in satellite imagery (Morison et al. 1987). Water temperature in the well mixed layer rose from near freezing to well over a degree above freezing as we crossed the front early on year day 191 (Fig. 5.8a). In Fig. 5.8, boxes mark two 6-h average blocks of data for which the friction velocity at 7 m was nearly identical, but the temperature and heat flux were quite different. Average w spectra for the two cases (Fig. 5.9) are similar in shape and magnitude except that for the melting (warm) regime there is a significant shift toward higher wave number. As shown, this implies that λ_{peak} is smaller, with the ratio being about 0.6. The vertical structure of both friction velocity and λ_{peak} for the two samples is illustrated in Fig. 5.10.

This is, of course, an almost singular sample, yet it provides a very useful template for looking at turbulence scales. The mean value of turbulent heat flux at 7 m for 6 h centered about time 191.125 is about 475 W m^{-2}, but this is not *a priori* representative of the basal heat. That can be estimated, however, from the "three-equation solution" for interface flux quantities discussed in detail in Chapter 6, using measured T, S, and u_*, providing a (water equivalent) basal melt rate of about 2.5 cm over the 6-h period from heat flux at the interface of about 350 W m^{-2}. This, in turn, implies a buoyancy flux $\langle w'b' \rangle_0 \approx 2 \times 10^{-7}$ m^2 s^{-3}, and from (4.26), $\eta_* \approx 0.65$. Since the planetary scale (u_{*0}/f) in each case is about the same, the ratio of scales in the outer layer should go as η_*.

An independent estimate of buoyancy flux may be made from the measured u_* and λ_{peak} by considering the TKE production ($P_s = u_*^3/\lambda_{peak}$) and TKE dissipation rate calculate via (5.3), as graphed in Fig. 5.11. As discussed below, we often find that in the outer layer, shear production exceeds dissipation. We found, for

---

[1] This does not pertain to dynamics in the outer part of the IOBL where buoyancy associated with quite modest melt rates can significantly limit the depth to which turbulent mixing penetrates.

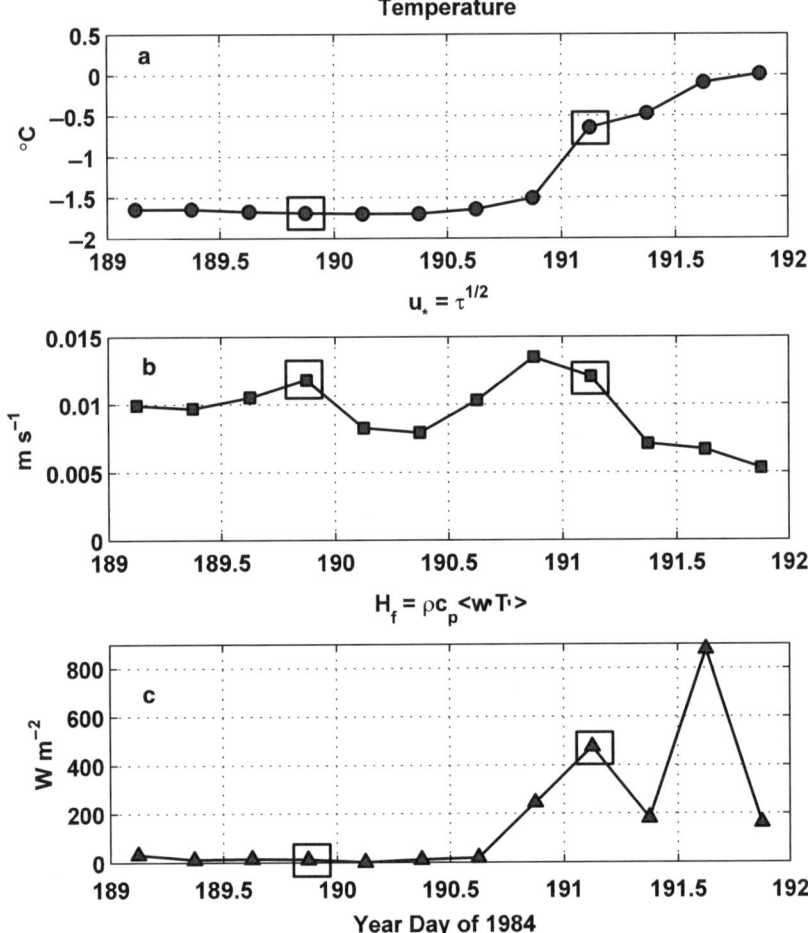

Fig. 5.8 Temperature **a**, salinity **b**, and friction velocity **c** measured 7 m below the ice during the MIZEX 1984 project in the Greenland Sea marginal ice zone. Data have been averaged in 6-h blocks (Adapted from McPhee 1994. With permission American Meteorological Society)

example, that for the values calculated at time 189.875, when the heat flux and melt rate were low enough to assume neutral stability, the ratio $P_s/\varepsilon \approx 1.67$. If we assume that the same ratio would hold at time 191.125 if there were no buoyancy flux, the expected value for P_s would be as indicated by the asterisk in Fig. 5.11. If the difference between the observed value and the neutral estimate represents the buoyancy sink ($P_b = -\langle w'b'\rangle \approx -0.22\,\mu\mathrm{W\,kg^{-1}}$) its value corresponds reasonably well to the estimated melt rate and buoyancy flux derived above, based on mixed layer characteristics.

5.2 The IOBL with Stabilizing Boundary Buoyancy Flux

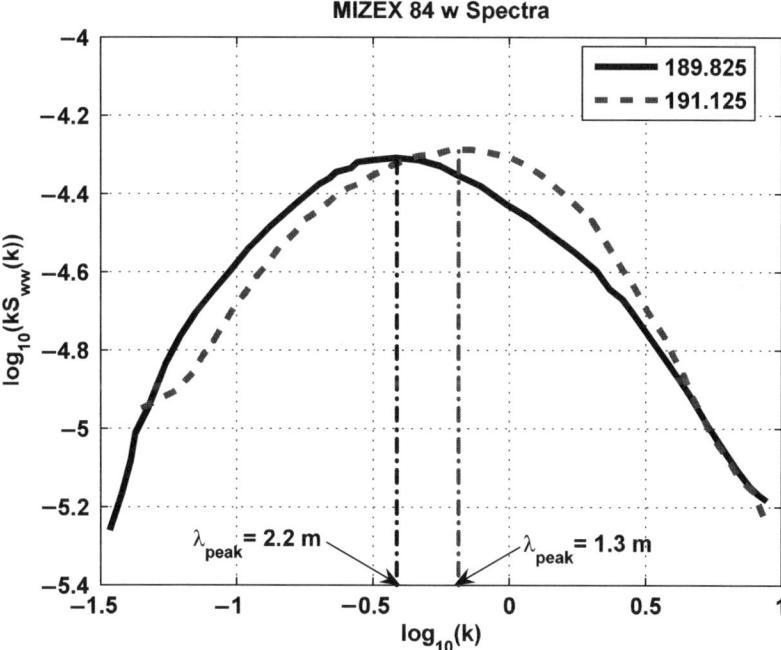

Fig. 5.9 Comparison of w spectra before (189.875) and after (191.125) crossing the temperature front. Spectra were analyzed in the same way as for Figs. 3.9 and 5.6, with $c_\lambda = 0.85$ (Adapted from McPhee 1994. With permission American Meteorological Society)

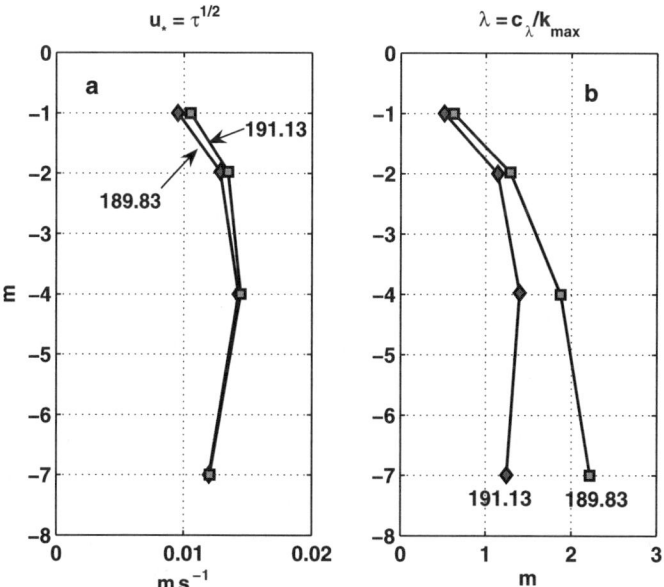

Fig. 5.10 a Comparison of friction velocity in cold water (189.83) and warm (191.13) at four levels during MIZEX 84. b. Corresponding turbulence length scales derived from peaks in the w spectra (Adapted from McPhee 1994. With permission American Meteorological Society)

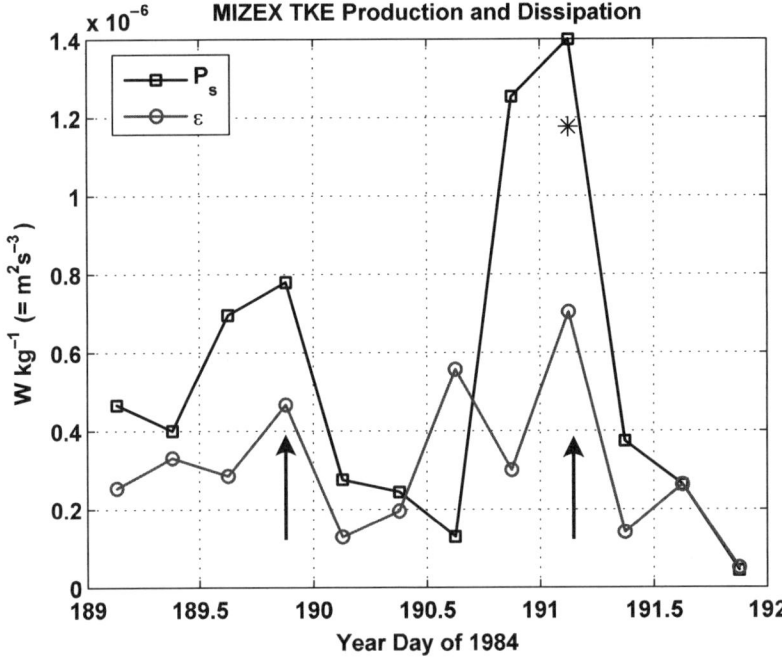

Fig. 5.11 MIZEX TKE production and dissipation estimates. Arrows indicate samples with similar friction velocity before and after crossing the thermal front. Asterisk at time 191.125 is the expected neutral value of P_s (Adapted from McPhee 1994. With permission American Meteorological Society)

5.3 The Statically Unstable IOBL

For the same reasons that it is difficult to easily measure conditions with rapid melting and stabilizing buoyancy flux, the statically unstable IOBL that persists when ice is freezing rapidly has been little studied. By far the most ambitious project aimed at examining the impact of freezing on boundary layer dynamics was the Lead Experiment (LeadEx) in March and April 1992. The strategy for LeadEx was to establish a base camp over the deep Canadian Basin north of Alaska, then wait for leads to open nearby. Once a suitable lead opened, two helicopters would transport equipment and shelters to the lead edge as rapidly as possible in order to measure as many characteristics of the freezing process and its impact on both atmosphere and ocean dynamics as possible. With temperatures typical of the Arctic in March and early April, leads do not generally remain open for long, so speed was important.

A scientific party comprising considerable collective experience was assembled under the leadership of Chief Scientist Jamie Morison, the base camp was established, and we waited, and waited. For many of us who had experienced the breakup of several ice stations (e.g., AIDJEX Big Bear and FRAM I) over the years, it seemed high irony that when we were looking and hoping for leads, Nature would

5.3 The Statically Unstable IOBL 99

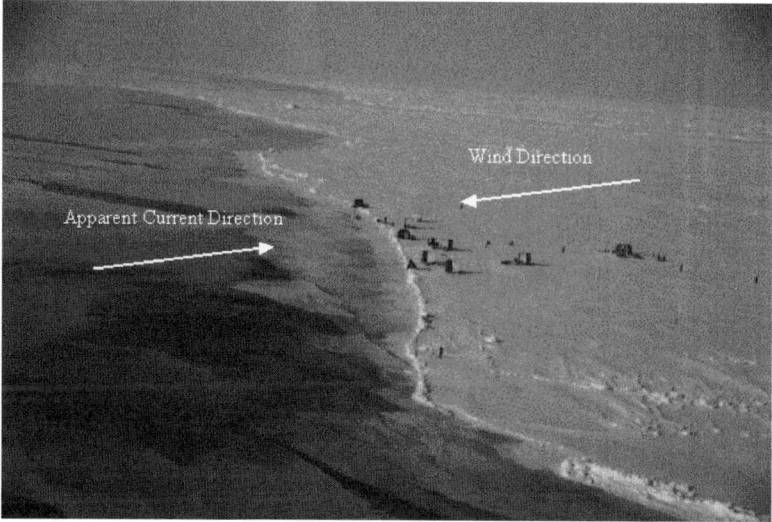

Fig. 5.12 Aerial view of LEADEX Lead 3 showing the temporary station and directions of the prevailing wind and apparent current

keep ice in our operating area just about as compact as possible. But finally, she relented and a kilometer-wide lead opened about 20 km south of the base camp (Fig. 5.12). In many ways this lead was almost ideal, in that its opening coincided with a moderate breeze from the north, meaning that our oceanographic instruments deployed on the northern "beach" sensed a boundary layer that appeared to be advecting toward us from the full fetch of the open water/thin ice region extending perhaps 30 boundary-layer thicknesses upstream.

Results reported in McPhee (1994), McPhee and Stanton (1996), and Morison and McPhee (1998) summarize oceanographic aspects of LeadEx. Those particularly focused on turbulence scales are reviewed here. A novel aspect of the TIC system deployed during LeadEx was the incorporation of a Sea-Bird Electronics SBE07 microstructure conductivity (μC) sensor into one of the instrument clusters. Calibration of the open-electrode μC sensors is much less stable than the standard SBE04 conductivity meters, which route the sampled water through an annular duct; however, their response to small salinity variations in the flow is superior to the unpumped SBE4 sensors. The μC equipped TIC provided credible estimates of salinity flux, hence direct measurements of buoyancy flux (almost entirely dependent on salinity) for the first time.

Turbulence measurements at the edge of lead 3 over about half a day (Fig. 5.13) show relatively mild stress with u_* about 7 mm s^{-1} and buoyancy flux of about -0.08μW kg^{-1}, which is somewhat less than half in magnitude of the stabilizing buoyancy flux described above in the MIZEX melting example. The mean Obukhov length was about -12 m with $\mu_* \approx -4$. The mean salinity flux over the period shown in Fig. 5.13 suggests that salt was entering the upper ocean from basal freezing at

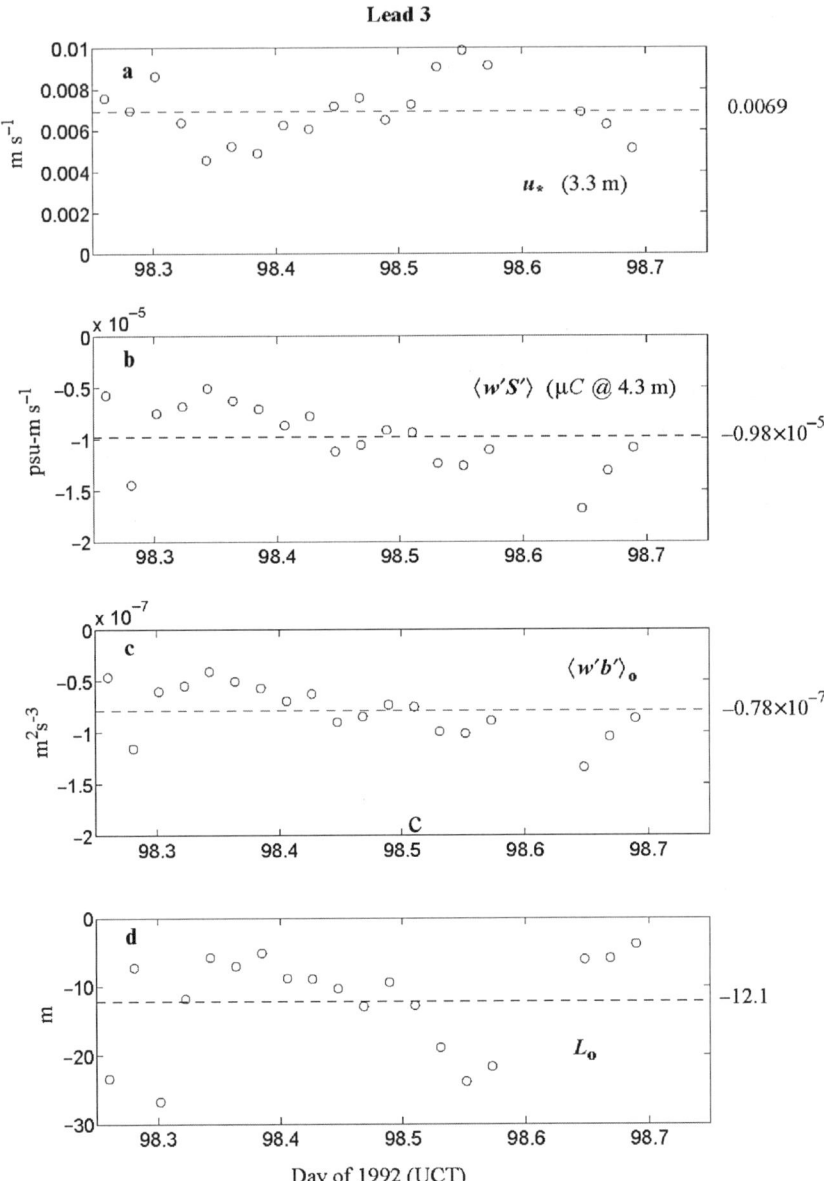

Fig. 5.13 a u_* at 3.3 m depth, LeadEx lead 3 at the edge of a freezing lead. **b** Salinity flux from TIC at 4.3 m equipped with a μC conductivity probe. **c** Estimated buoyancy flux. **d** Obukhov length (From McPhee and Stanton 1996. With permission American Geophysical Union)

a rate of roughly 4 cm per day. Air temperature during the lead 3 deployment remained cold (-20 to -30°C). So even with relatively mild stress (0.05 Pa) and cold temperature, the magnitude of u_* remained small compared with values seen in midlatitude atmospheric studies. For context, if scales in the atmosphere are roughly

5.3 The Statically Unstable IOBL

30 times greater than in the IOBL, the corresponding dimensionless height at the standard 10-m sampling level in the atmosphere would be $\zeta = z/L \simeq -0.03$, i.e., from an atmospheric surface layer perspective, the surface layer under the freezing lead remains close to neutral stability.

During the lead 3 deployment, we used a mast with four TICs spanning 6 m that could be lowered as a unit in the upper ocean. During most of the time (~ 12 h) the mast was positioned within the upper 10 m of the water column, but was lowered to about midway in the mixed layer (14–20 m) for the last 3 h with sufficient relative current for acceptable turbulence measurements (McPhee and Stanton 1996).

By again considering peaks in the area-preserving w spectra, we calculated average λ_{peak} values for the various measurement levels (Fig. 5.14). The significant feature was that the scales inferred from the spectral peaks were almost an order of magnitude greater than would be expected had the IOBL been neutrally stratified, i.e., $\lambda_N = \Lambda_* u_{*0}/f \simeq 1.4$ m. The curves plotted in Fig. 5.14 represent the mixing lengths associated with the unstable dimensionless shear formulations discussed in Section 4.1.3. The vertical dashed line, which closely matches λ_{peak} at 9.4 m depth, is κd_{ml}, i.e., von Kármán's constant times the depth of the well mixed layer, 28 m during lead 3.

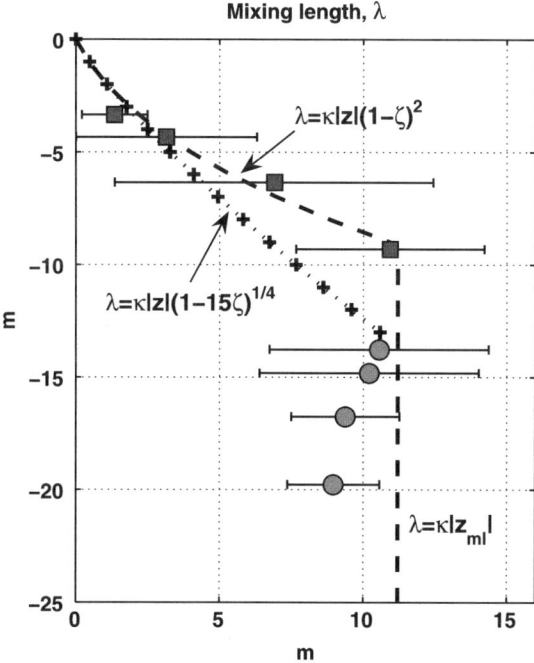

Fig. 5.14 Average mixing length from spectral peaks when the mast was shallow (square symbols) and mid-depth in the well mixed layer (circles). A limit based on κ times the depth of the well mixed layer is indicated by the dashed vertical line (Adapted from McPhee 1994. With permission American Meteorological Society)

For LeadEx, the turbulence mast measurements were complemented by a loose-tethered microstructure profiler (LMP) that provided high-resolution profiles of temperature, conductivity, and small-scale velocity shear (McPhee and Stanton 1996). Turbulent heat flux was thus measured in two ways during LeadEx: by direct covariance at four TIC levels, and indirectly by estimating thermal dissipation rate from

$$\varepsilon_T = 3\nu_T \left\langle \frac{\partial T}{\partial z} \right\rangle^2$$

assuming isotropy as well as local balance between thermal variance production and dissipation, viz.

$$\langle w'T' \rangle_{LMP} = -\varepsilon_T / \frac{\partial T}{\partial z}$$

We also estimated mean temperature gradient in the well mixed layer by two methods. First, following earlier work, we applied small constant corrections to the TIC thermometers during periods of near zero heat flux in order to eliminate apparent thermal gradients during a time when we expected the actual potential temperature gradient to vanish (i.e., we again used the polar mixed layer as a calibration bath). We then calculated thermal gradient by least-squares fitting of the four TIC thermometers. Thermal gradient was also calculated by least-squares fitting of 1-h average LMP potential temperature profiles in the depth range from 3 to 20 m, where visual inspection of the profiles indicated linear dependence. Results for lead 3 (Fig. 5.15, from McPhee and Stanton 1996) show that near solar noon, as much as 10% of the incoming solar radiation had penetrated the thin (5–10 cm) ice cover in the lead to be mixed downward by turbulence. At night, heat was extracted from the upper ocean.

With measurements of heat flux and thermal gradient, we estimated the thermal eddy diffusivity as regression slopes of kinematic heat flux versus negative thermal gradient (Fig. 5.16). Although error bars are large ($K_{T(LMP)} = 0.0443 \pm 0.0092$; $K_{T(TIC)} = 0.0473 \pm 0.0214$; McPhee and Stanton 1996) the estimates are consistent with each other and with large eddy viscosity. The thermal mixing length λ_T is about 7 m, several times larger than that estimated for a neutral layer with higher surface stress (Fig. 5.2b).

5.4 Velocity Scales in the IOBL

So far in the development of both the similarity theory and the scaling based on vertical velocity spectra, we have used only one velocity scale, u_*, for the IOBL. Others are possible, and perhaps preferable in some situations. In many of the early studies of boundary-layer flow, the "free-stream" current speed was used as the scale velocity,[2] and for a strictly quadratic drag relationship (which is often appropriate

---

[2] In the case of drifting ice, the "free-stream" velocity would be the ice velocity relative to geostrophic flow in the ocean, which is often relatively small. It is analogous exactly to the geostrophic wind in the atmospheric boundary layer.

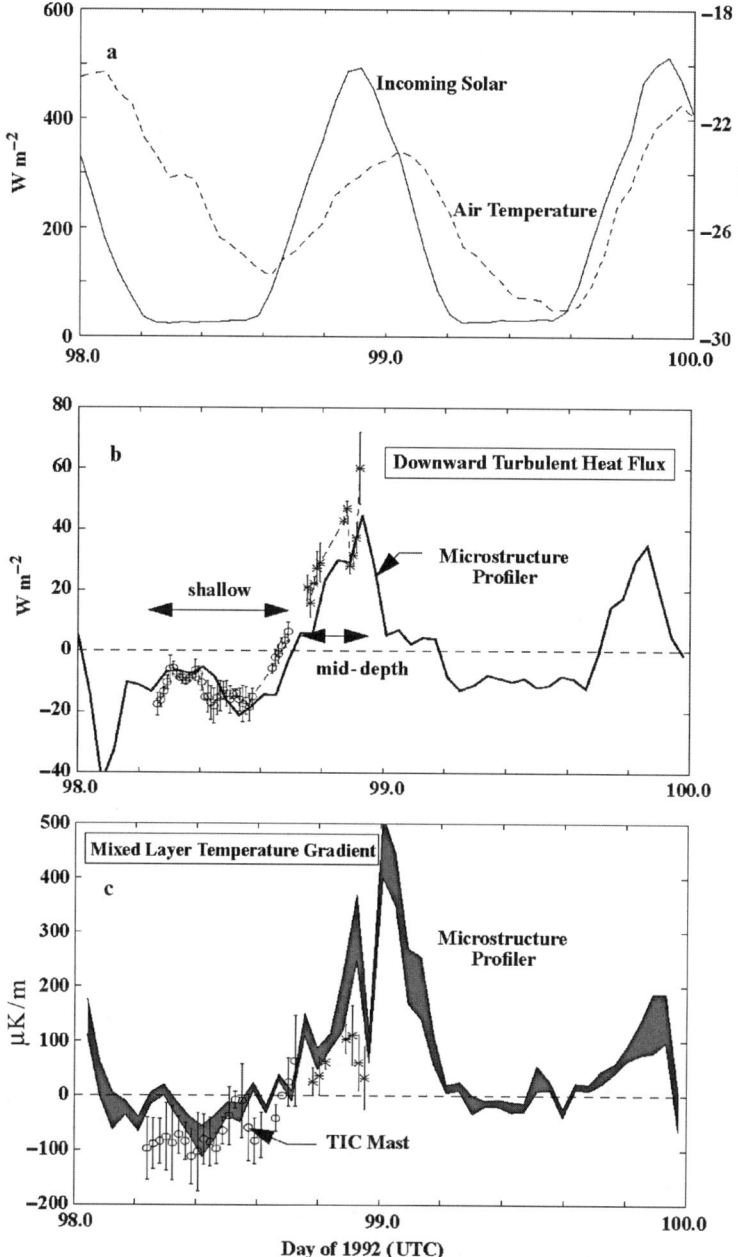

Fig. 5.15 a Incoming shortwave radiation measured at the surface and air temperature at LeadEx lead 3 site. **b** *Downward* turbulent heat flux, average over the four TICs (x symbols with std deviation error bars), and from LMP microstructure measurements in the depth interval from 7.5 to 12.5 m. **c** Temperature gradient from the TIC mast (circles with error bars indicating the 95% confidence interval for the linear regressions), and from the LMP mean temperature gradients from 3 to 20 m depth (shaded curve) (From McPhee and Stanton 1996. With permission American Geophysical Union)

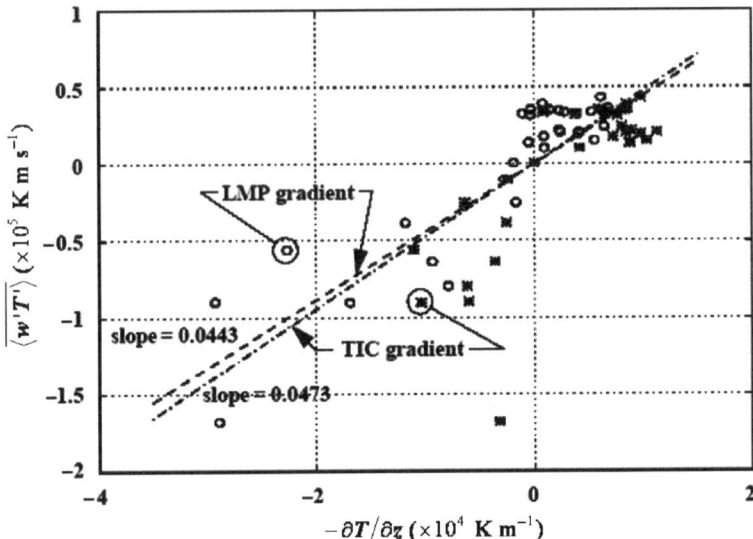

Fig. 5.16 Scatter plot of $\langle w'T'\rangle$ vs. thermal gradient for LeadEx lead 3, from LMP data (circles) and TIC mast (asterisks). Slopes represent two estimates of eddy thermal diffusivity in the upper ocean under the lead

for the neutral surface layer), this would suffice since $u_*\propto$ to current speed. However, as illustrated by Fig. 4.5, when rotation comes into play, V_s is probably a poor choice since it is highly dependent on z_0 even if the underlying outer layer remains unchanged.

Another scale velocity often suggested (e.g., Mellor and Yamada 1982) is the square root of the trace of the Reynolds stress tensor (equivalent to the square root of twice the TKE per unit mass)

$$q = \langle u_i' u_i'\rangle^{1/2} \qquad (5.7)$$

Most of our studies of the Reynolds stress tensor have shown a fairly tight proportionality relationship between u_* and q. A subtle factor from the observational standpoint is that because turbulence transports properties mainly at the scales of the "energy-containing" eddies, the covariance provides an effective filter for time series with high frequency content not related to turbulence. This is sometimes important, for instance, with acoustic backscatter current meters, which tend to be electronically "noisy" when operating in a fluid with sparse sound scatterers, often typical of polar mixed layers. In that situation we find that u_* based on covariance statistics is relatively unaffected by high-frequency instrumental noise, whereas q obviously is. From a modeling standpoint, using q as the turbulent scale velocity also means carrying a conservation equation for TKE in the model solution.

The problem with u_* as a scale velocity comes when an important source of turbulence in the flow is destabilizing buoyancy flux in addition to shear. In the case of pure free convection (where shear plays no role at all), from dimensional considerations a scale velocity is

$$w_{*F} \propto \left(-\lambda_{max}\langle w'b'\rangle\right)^{1/3}$$

For forced convection, where both shear and buoyancy driven convection are present, as in the LeadEx example above, we postulate a mixed scale velocity:

$$w_* = \left(u_*^3 - \lambda_{max}\langle w'b'\rangle\right)^{1/3}; \qquad \langle w'b'\rangle < 0 \qquad (5.8)$$

Using average values shown if Fig. 5.14, $w_* \approx 0.01\,\mathrm{m\,s^{-1}}$ versus $u_* \approx 0.007\,\mathrm{m\,s^{-1}}$. The example illustrates again that even with fairly rapid freezing, the impact of buoyancy-forced convection in the IOBL is proportionately much less important than convection in the atmospheric boundary layer over a heated land surface.

5.5 Summary of IOBL Scales

From turbulence studies in the IOBL exemplified by the examples described above, a heuristic view of turbulence scales is presented in Fig. 5.17. When melting or freezing is negligible compared with shear production of TKE (Fig. 5.17a),

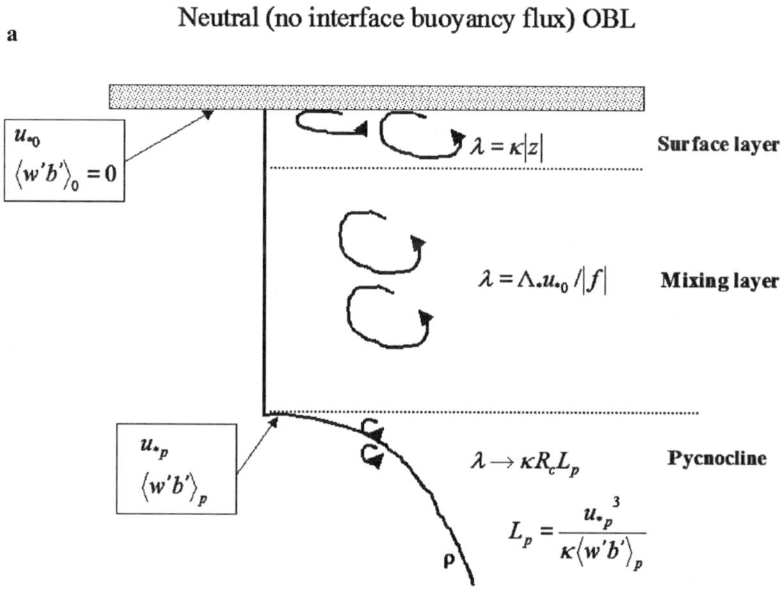

Fig. 5.17 Schematic illustrating postulated scales for the IOBL under varying surface flux (momentum and buoyancy) conditions. **a** Neutral stability in the well mixed layer. **b** Stabilizing buoyancy flux from melting. **c** Destabilizing buoyancy flux from freezing

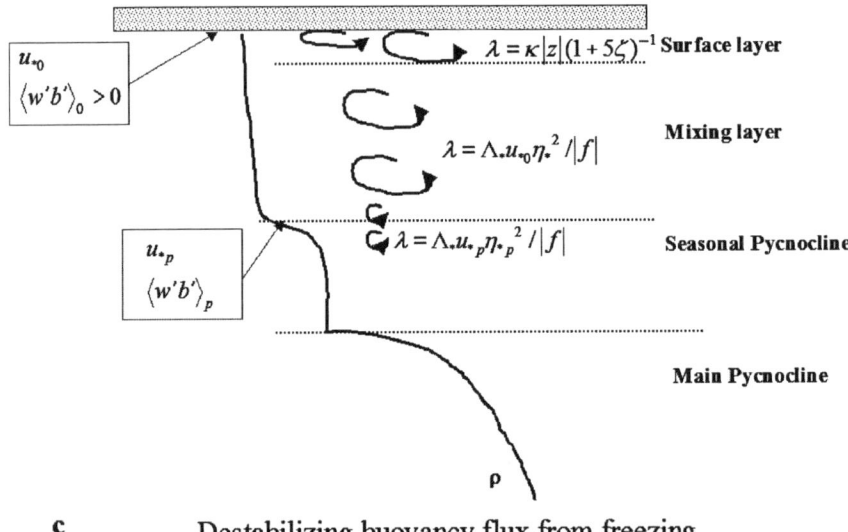

Fig. 5.17 Continued

λ increases with distance from the boundary until it reaches a limiting value determined by the dynamic planetary depth, which typifies eddies in the remainder of the well mixed layer. Any changes in density structure will come from mixing in the upper pycnocline, with scales u_{*p} (the square root of turbulent stress magnitude at z_p) and with $\lambda_p = \eta_{*p}^2 u_{*p}/|f|$. The idea here is that an Ekman layer will form in the upper pycnocline, forced by the stress and buoyancy flux ($\langle w'b'\rangle_p$) at the interface between the well mixed layer and the stratified fluid below. For typical Arctic density structure, the pycnocline is strong enough that

$$\eta_{*p}^2 = \left(1 + \frac{\Lambda_* \langle w'b'\rangle_p}{|f|R_c u_{*p}^2}\right)^{-1} \rightarrow \frac{|f|R_c u_{*p}^2}{\Lambda_* \langle w'b'\rangle_p}$$

and $\lambda_p \rightarrow \kappa R_c L_p$.

When melting occurs following a period when the well mixed layer is relatively deep, a seasonal pycnocline forms as in Fig. 5.17b. The depth of this intermediate layer (z_p) depends on the turbulence scales determined by the surface flux conditions. If this layer is shallow and the newly formed seasonal pycnocline is weak, then η_{*p}^2 may be close to unity, in which case turbulent exchange in the Ekman layer formed below z_p will be similar to what it would be without the intermediate pycnocline. In the surface layer, λ follows Monin-Obukhov similarity (although for realistic values of μ_* the dimensionless shear within the surface layer departs from 1 by only a few percent.)

When freezing injects salt into the IOBL, turbulence scales are enhanced and the scale of the energy-containing eddies will approach a significant fraction (c_{ml}) of the depth of the well mixed layer (Fig. 5.17c). Consequently, λ increases with distance from the boundary following a Monin-Obukhov function ($\lambda = \kappa |z|(1-\zeta)^2$) until it reaches a value $c_{ml}|z_p|$, which it retains until the pycnocline is encountered. LeadEx results indicated that $c_{ml} \sim \kappa$.

References

Busch, N. E. and Panofsky, H. A.: Recent spectra of atmospheric turbulence. Quart. J. R. Met. Soc., 94, 132–147 (1968)

Deardorff, J. W.: Numerical investigation of neutral and unstable planetary boundary layers. J. Atmos. Sci., 29, 91–115 (1972)

Edson, J. B., Fairall, C. W., Mestayer, P. G., and Larsen, S. E.: A study of the inertial-dissipation method for computing air-sea fluxes. J. Geophys. Res., 96, 10,689–10,711 (1991)

Hinze, J. O.: Turbulence, Second Edition. McGraw-Hill, New York (1975)

McPhee, M. G.: On the turbulent mixing length in the oceanic boundary layer. J. Phys. Oceanogr., 24, 2014–2031 (1994)

McPhee, M. G.: Physics of early summer ice/ocean exchanges in the western Weddell Sea during ISPOL, Deep-Sea Res., II, doi:10.1016/j.dsr2.2007.12.022, in press (2008)

McPhee, M. G. and Martinson, D. G.: Turbulent mixing under drifting pack ice in the Weddell Sea. Science, 263, 218–221 (1994)

McPhee, M. G. and Smith, J. D.: Measurements of the turbulent boundary layer under pack ice. J. Phys. Oceanogr., 6, 696–711 (1976)

McPhee, M. G. and Stanton, T. P.: Turbulence in the statically unstable oceanic boundary layer under arctic leads. J. Geophys. Res., 101, 6409–6428 (1996)

Morison, J. H., McPhee, M. G., and Maykut, G. A.: Boundary layer, upper ocean, and ice observations in the Greenland Sea marginal ice zone. J. Geophys. Res., 92, 6987–7011 (1987)

Morison, J. H. and McPhee, M. G.: Lead convection measured with and autonomous underwater vehicle. J. Geophys. Res., 103, 3257–3281 (1998)

Chapter 6
The Ice/Ocean Interface

Abstract: At first glance it seems that describing heat exchange at the ice/ocean interface would be a reasonably straightforward exercise in applying the first law of thermodynamics. There are essentially only three important factors in the enthalpy balance: (i) upward (or downward) heat conduction within the ice column; (ii) heat flux from or to the underlying ocean; and (iii) latent heat associated with the phase change as ice grows or melts. However, salt greatly complicates the process. In the same way that spreading salt may help remove ice from a cold roadway, diffusion of salt from the ocean lowers the freezing point at the interface so that whenever the IOBL temperature is above freezing, upward heat transfer occurs. Melting occurs when upward heat flux from the ocean exceeds upward conduction in the ice column; freezing happens when the ice conduction exceeds ocean heat flux (which is why ice can and does form even when the upper ocean is warmer than its salinity-determined freezing temperature).

In the initial efforts at modeling sea ice, the prescribed ocean heat flux was important in maintaining the modeled equilibrium mass balance. Maykut and Untersteiner (1971), for example, found that a constant 2 W m^{-2} heat flux from the ocean was required for a realistic equilibrium thickness of Arctic pack ice. For Southern Ocean sea ice, Parkinson and Washington (1979) also utilized a constant ocean heat flux in their model, but found that it needed to be an order of magnitude greater—about 25 W m^{-2}. If the early models considered the ocean mixed layer at all, it was assumed that if sea ice was present, the mixed layer would remain at its freezing temperature, which in essence meant that any heat entering the upper ocean, either by absorption of solar radiation or upward conduction from below, would be instantaneously transferred to the ice.

When summer measurements of IOBL characteristics with modern instrumentation became available, it was obvious that the polar mixed layers could remain above freezing for extended periods. A smoothed time series of mixed-layer temperature at AIDJEX station Blue Fox, on the eastern side of the Beaufort Gyre over the Canadian Basin illustrates (Fig. 6.1) that even in the central Arctic with perennially high ice concentrations, a large amount of heat is stored in the upper ocean for at least a third of the year. The time series is more or less typical of the other AIDJEX

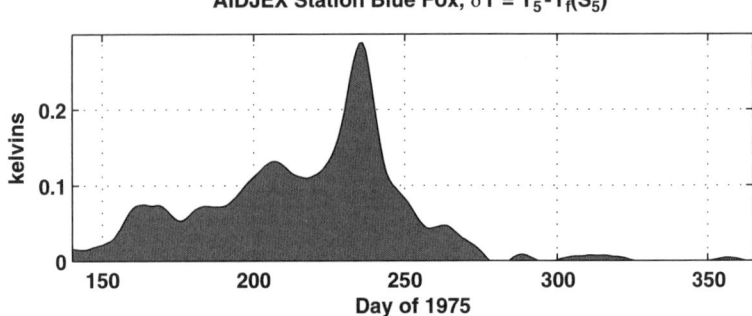

Fig. 6.1 Elevation of near surface (5 m from ice) temperature in the water column above its freezing temperature (at surface pressure) at Station Blue Fox during the 1975–1976 AIDJEX experiment

stations, the SHEBA data in 1998, and several unmanned buoy drifts covering most of the Arctic Ocean. In this chapter we will explore how this storage comes about and what it implies about survivability of sea ice.

6.1 Enthalpy and Salt Balance at the Interface

The thermal balance at the interface may be represented in terms of a control volume following the interface (Fig. 6.2). The interface velocity, w, comprises two terms: $w_0 = -\frac{\rho_{ice}}{\rho_w}\dot{h}$ is the isostatically adjusted bottom melt rate where $\dot{h}$ is ice growth rate; and w_p is a "percolation velocity," expressing the idea that even if there were no bottom melting, a hydrostatic head at the surface might force meltwater through a porous ice cover, resulting in a basal velocity. Ignoring sensible heat associated with the percolation velocity,[1] the idealized control volume reflects primarily a balance among (i) conduction in the ice column; (ii) turbulent heat flux from the ocean; and (iii) latent heat release or absorption from melting or freezing. Consequently a simple expression for the "kinematic" interface heat balance is

$$-\dot{q} + \langle w'T' \rangle_0 = w_0 Q_L \qquad (6.1)$$

where "kinematic" refers to the actual heat equation divided by the product of density and specific heat[2] $(\rho_w c_p)$, with units K m s^{-1}. Note that if the sum of terms on the left hand side is negative (ice conduction exceeds ocean heat flux), w_0 is negative indicating freezing.

---

[1] In a situation where w_p was important, the temperature gradient near the interface would be very small.

[2] At surface pressure the specific heat of seawater is a weak function of temperature and salinity (see, e.g., Appendix A3.4 of Gill 1982). For typical Arctic mixed layer properties, it is very close to $4,000$ J kg^{-1} K^{-1}.

6.1 Enthalpy and Salt Balance at the Interface

Thermal Balance at the Ice/Ocean Interface

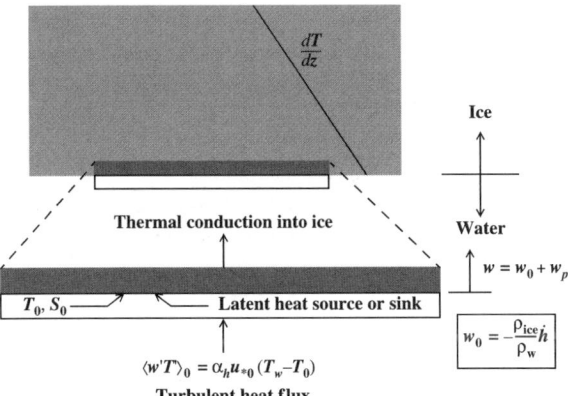

Fig. 6.2 Schematic of vertical exchanges at the ice/ocean interface in a small control volume (detail) following the ice/ocean interface

The kinematic ice conduction term is

$$\dot{q} = \frac{-K_{ice}T_z}{\rho_w c_p} \qquad (6.2)$$

where T_z is the thermal gradient near the base of the ice column, and K_{ice} is thermal conductivity of sea ice there. The latter is a fairly complicated function of brine volume and temperature, but here we use a simplified approximation suggested by Untersteiner (1961):

$$K_{ice} \approx K_{fresh} + \beta S_{ice}/T_{ice}$$

where S_{ice} and T_{ice} are ice salinity (practical salinity scale) and temperature (°C) averaged over that part of the ice column spanned by T_z; K_{fresh} is thermal conductivity for fresh ice (2.04 J m^{-1} K^{-1} s^{-1}) and β is a constant (0.117 J m^{-1} K^{-1} s^{-1} psu^{-1} °C).

The effective (apparent) latent heat of sea ice is also strongly affected by variation in brine volume and temperature. For the range of ice temperatures near the base of the ice column, a rough approximation the formulas for the brine volume and latent heat given by Maykut (1985, his equations (6.13) and (19)) is

$$Q_L = \frac{L_{ice}}{c_p} = \frac{L_{fresh}}{c_p}(1 - 0.03 S_{ice})$$

where Q_L is a "latent heat temperature" with units kelvins, and $L_{fresh} = 333.5$ kJ kg^{-1} is latent heat for fresh ice. Both approximations for the ice properties are crude, but should be considered in light of large uncertainties in ice salinity and brine volume near the interface, particularly if ice is freezing.

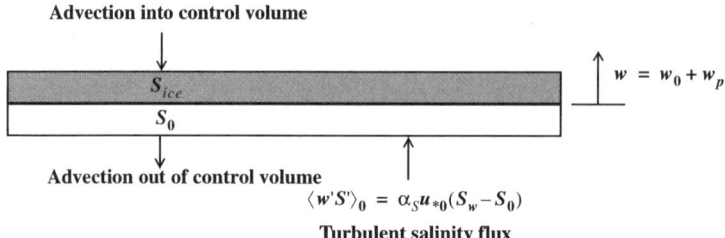

Fig. 6.3 As in Fig. 6.2 except salt balance at the ice/ocean interface

Since there is no counterpart in the interface salt balance to latent heat exchange, a similar control volume approach (Fig. 6.3) yields a balance between turbulent salinity flux and the advection of relatively fresh meltwater entering the control volume at the top (for positive w) and water with interface salinity (S_0) leaving at the bottom:

$$\langle w'S' \rangle_0 + w(S_{ice} - S_0) = 0 \qquad (6.3)$$

In this case $w = w_0 + w_p$ represents the total interface vertical velocity, thus allows a salinity flux associated with percolation, even in the absence of melting or freezing.

6.2 Turbulent Exchange Coefficients

It is customary in considering parameterization of turbulent exchanges to express the flux of some quantity at the boundary as the product of a scale velocity and the change in the quantity from the boundary to some reference level. For example, in the surface layer discussed in Chapter 5, momentum flux may be parameterized in terms of an exchange coefficient α_m as

$$\tau_0 = u_{*_0}^2 = \alpha_m u_{*_0} \Delta U$$

where ΔU is the fluid speed measured at some distance d from the boundary (since surface velocity is zero). Thus

$$\alpha_m = \frac{u_{*_0}}{\Delta U} = \left(\frac{1}{\kappa} \log \frac{d}{z_0} \right)^{-1} = \sqrt{c_D}$$

where c_D is a conventional quadratic drag coefficient referenced to d. For multiyear pack ice a representative value of z_0 is 0.05 m, and if we choose d in the surface layer, say 2 m, $\alpha_m = 0.11$.

6.2 Turbulent Exchange Coefficients

Analogous expressions for turbulent heat and salinity flux are

$$\langle w'T'\rangle_0 = \alpha_h u_{*0} \delta T \qquad (6.4)$$
$$\langle w'S'\rangle_0 = \alpha_S u_{*0} \delta S$$

where $\delta T = T_w - T_0$ and $\delta S = S_w - S_0$ are the differences in temperature and salinity between the far-field and interface values. If we assume that the analog holds exactly so that scalar and momentum exchange coefficients are comparable (a variant on Reynolds analogy), then it becomes clear why early ice/ocean models (e.g., Josberger 1983; Ikeda 1986) assumed that mixed-layer temperature remained near freezing. Suppose that $\alpha_h \approx \alpha_m = 0.11$, $u_{*0} = 0.01$ m s^{-1}, and $\delta T = 1$ K. During the 1984 MIZEX experiment in the marginal ice zone of the Greenland Sea, similar conditions persisted for at least a day as our floe drifted south over an ocean front (see Fig. 5.8 and McPhee et al. 1987). With $S_{ice} = 3$ psu, and $\dot{q} = 0$, this would imply a melt rate of about 1.3 m per day; however, we observed only about 7 cm of bottom ablation (corroborated by direct turbulent heat flux measurements). It is worth recounting that prior to MIZEX we had considered that the exchange coefficients, couched in terms of reduced roughness lengths for scalar variables by analogy with atmospheric studies, would be much smaller than for momentum (Mellor et al. 1986); nevertheless, when we drifted over warm water during the later part of MIZEX, the observed melt rate and ocean heat flux were much less than we anticipated. In retrospect, the fact that our floe survived when theory suggested it should have melted began the quest for a proper description of the exchange coefficients for heat and salt that still continues.

If the exchange coefficients for scalar quantities are different from momentum, dimensional analysis (e.g., Barenblatt 1996) may help in suggesting the functional form. Suppose we choose as our scalar quantity the kinematic heat flux and postulate that it depends on the following quantities:

$$\langle w'T'\rangle_0 = F(\delta T, u_{*0}, v_T, v, z_0)$$

Then by the Pi Theorem, there are five governing parameters, three with independent dimensions, so that

$$\frac{\langle w'T'\rangle_0}{u_{*0} \delta T} = \alpha_h \left(\frac{v}{v_T}, \frac{u_{*0} z_0}{v}\right) = \alpha_h \left(\mathrm{Pr}, \mathrm{Re}_*\right) \qquad (6.5)$$

where Pr and Re_* are the Prandtl and surface friction Reynolds numbers, respectively. A similar expression for salinity flux is

$$\frac{\langle w'S'\rangle_0}{u_{*0} \delta S} = \alpha_S \left(\frac{v}{v_S}, \frac{u_{*0} z_0}{v}\right) = \alpha_S \left(\mathrm{Sc}, \mathrm{Re}_*\right) \qquad (6.6)$$

where Sc is the Schmidt number. The Prandtl and Schmidt number dependency suggests that for scalar quantities, molecular diffusivities may be important and indeed this turns out to be the case. Note that α_h and α_S represent the inverse of dimensionless changes in temperature and salinity across the boundary layer.

A starting point for evaluating the functional form of the dimensionless variables in (6.5) and (6.6) comes from the Blasius solution for momentum and contaminant exchange in laminar flow over a flat plate (e.g., Incropera and DeWitt 1985). The analysis expresses the Stanton number (kinematic heat flux divided by the product of far-field velocity and δT) in terms of the Prandtl and Reynolds number:

$$St \propto \mathrm{Re}^{-p} \mathrm{Pr}^{-n} \qquad (6.7)$$

where the exponents are $p = 1/2$ and $n = 2/3$ for the laminar case.

Owen and Thomson (1963) and Yaglom and Kader (1974) adopted the functional form (6.7) in interpreting results from laboratory studies of mass transfer in turbulent flow over hydraulically rough surfaces. Each analysis considered a "transition sublayer" in which molecular effects dominated, greatly reducing the exchange coefficients for scalar quantities compared with momentum. Owen and Thomson treated the exponents as empirical constants and suggested that $p = 0.45$ and $n = 0.8$. Yaglom and Kader, on the other hand, assumed a form like (6.7), but with several small additive constants. For large Prandtl (Schmidt) numbers, their expression approaches (6.7) with $p = 1/2$ and $n = 2/3$. Both approaches assumed that the dimension of the transition sublayer (the length scale in Re) was the same as the height of the roughness elements in the laboratory flows, which is typically $30 \times z_0$. McPhee et al. (1987) demonstrated that the value for α_h inferred from direct measurements of heat flux, friction velocity and far-field temperature and salinity was about 0.0055, on a day when mixed layer temperature was well above freezing during MIZEX (Section 5.2). This was consistent with (6.7) when the proportionality constant was 1.57, which is about double that recommended by Yaglom and Kader (1974). It was based on a relatively large value of Re reflecting large surface roughness in the marginal ice zone, which we pointed out was not necessarily appropriate when considering the transition sublayer.

6.3 The "Three-Equation" Interface Solution

Substituting the exchange coefficient parameterization (6.4) into (6.1) and (6.3), the conservation equations in the control volume are

$$-\dot{q} + \alpha_h u_{*0} (T_w - T_0) - Q_L w_0 = 0 \qquad (6.8)$$
$$\alpha_S u_{*0} (S_w - S_0) + w(S_{ice} - S_0) = 0$$

In general, we seek to establish how fast ice is melting or freezing (w_0), hence the interface scalar fluxes, $\langle w'T' \rangle_0$ and $\langle w'S' \rangle_0$, in terms of prescribed quantities which include u_{*0}, T_w, S_w, $\dot{q}$, S_{ice}, w_p and the exchange coefficients. In other words we seek a solution for three variables (T_0, S_0, w_0). This requires a third equation for closure, specified by assuming that the interface remains at its salinity determined freezing temperature: $T_0 = T_f(S_0) \approx -mS_0$ where m is the slope of the "freezing

6.3 The "Three-Equation" Interface Solution

line," determined from the UNESCO formula for freezing temperature as a function of T and S for values near the interface (e.g., Gill 1982). Combining this with (6.8) then provides a quadratic equation for, e.g., S_0, in terms of three temperature scales

$$mS_0^2 + (T_H + T_L - mS_{ice} + T_P)S_0 - (T_H + T_P)S_{ice} - T_L S_w = 0 \qquad (6.9)$$

where

$$T_H = T_w - \frac{\dot{q}}{\alpha_h u_{*_0}}; \qquad T_L = \frac{\alpha_S Q_L}{\alpha_h}; \qquad T_P = \frac{w_P Q_L}{\alpha_h u_{*_0}}$$

from which w_0 and T_0 follow, along with the interface fluxes. Normally the percolation velocity is small enough that the last scale is inconsequential.

Double diffusion is a term coined to describe what occurs when scalar contaminants in a fluid diffuse at different rates. In cold seawater, for example, molecular thermal diffusivity is about 200 times greater than salt diffusivity. So if one were to place a parcel of relatively warm, saline water next to a cold, fresh parcel in a quiescent setting, at some distance from the initial boundary, the change in temperature with time relative to the initial difference in temperature would be far greater than the change in salinity relative to its initial difference. In most oceanographic usage, double diffusion refers to the fact that since seawater density depends on both temperature and salt concentration, different diffusion rates can lead to density differences that will cause fluid motion and mixing, even in the absence of any external forces besides gravity. In a different context, double diffusion in the thin water layer adjacent to the ice/ocean boundary is potentially quite important for the heat and mass balance of sea ice, because the relatively small observed melt rates described above imply that the Prandtl and Schmidt numbers, which depend on molecular diffusivities, control the exchanges of heat and salt. We stress that the development here concentrates on diffusion across a thin fluid sublayer on the liquid side of the interface, as opposed to how heat and salt are diffused within the ice crystal lattice. Although perhaps not obvious at the outset, mechanisms for the latter appears to quite different for melting versus freezing (see, e.g., Feltham et al. 2006), and this turns out to have important consequences the exchange coefficients, as described below.

Note from (6.9) that if $\dot{q}$ is negligibly small, as it often is in the summer ice with rapid melting conditions, then the quadratic solution depends on the ratio of exchange coefficients $R = \alpha_h/\alpha_S$, which is a measure of the strength of double diffusion at the interface. If $R = 1$, there is no double diffusion and salt plays a passive role in the heat exchange. As R increases, heat transfer increases relative to salt transfer. Returning to the control volume artifice, if $R = 1$ then just enough salt will enter the control volume to keep $T_0 \approx T_f(S_w)$ and heat transfer will continue unabated. For $R \gg 1$, the downward flux of water freshened by melt is inhibited, so that $T_0 > T_f(S_w)$ and the thermal driving is reduced. In this way, melting is rate limited by double-diffusive effects in the transition sublayers for heat and salt.

6.4 Heat Flux Measurements and the Stanton Number for Sea Ice

The first direct measurements of heat flux in the IOBL were made during the 1984 MIZEX project in the marginal ice zone of the Greenland Sea using instrument clusters combining fast response thermometers and small mechanical current meters capable of measuring into the inertial subrange of the turbulent spectrum (McPhee et al. 1987). A summary of the MIZEX heat flux measurements (Fig. 6.4) illustrates that in general there are increasing trends along axes representing increasing mixed-layer temperature elevation above freezing, and increasing friction velocity. Note that there is a distinction between T_0 and mixed-layer freezing temperature, $T_f(S_{ml})$, since if there is melting at the interface, S_0 will be different from far-field (mixed-layer) salinity. Consequently we make a distinction between the definition of α_h and a bulk Stanton number

$$St_* = \frac{\langle w'T' \rangle_0}{u_{*0} \Delta T} \tag{6.10}$$

where $\Delta T = T_w - T_f(S_w)$. If we identify T_w with mixed (or mixing) layer temperature, ΔT is readily measured, while δT is not.

Heat flux measurements were been made during several ice station experiments following MIZEX. We have found on a number of occasions that solar radiation penetrating into the water column can influence local measurements, and that caution

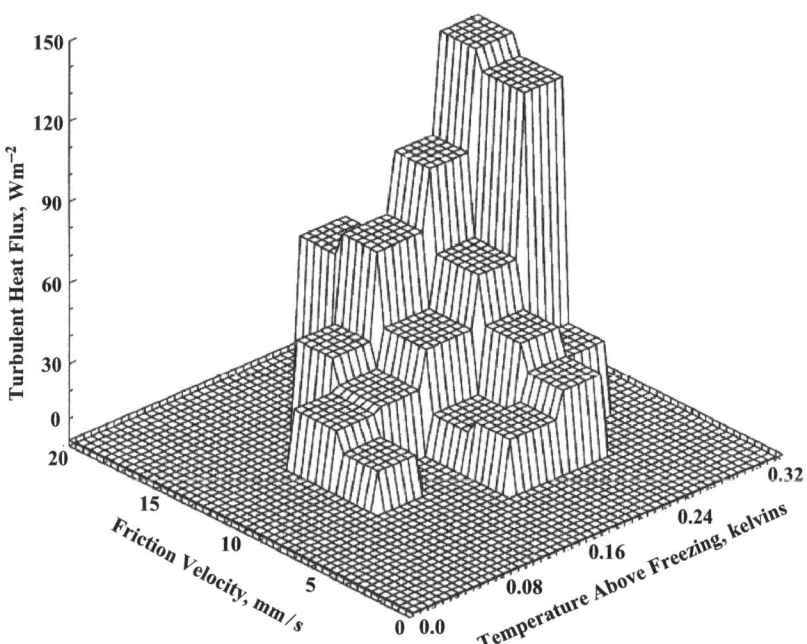

Fig. 6.4 Summary of directly measured heat flux and friction velocity averaged in bins for all data from MIZEX 84 (From Mcphee 1992. With permission American Geophysical Union)

6.4 Heat Flux Measurements and the Stanton Number for Sea Ice

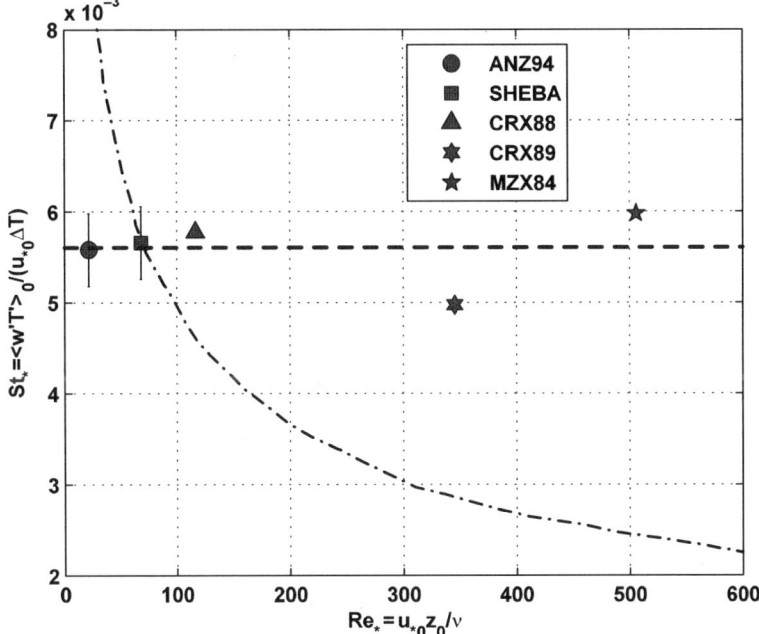

Fig. 6.5 Average values of bulk heat transfer coefficients St_* versus mean surface friction Reynolds number Re_* for five different ice drift projects, MIZEX 1984 (Greenland Sea marginal ice zone, summer), CEAREX 1988 (eastern Arctic Ocean, fall), CEAREX 1989 (north of Fram Strait, late winter), ANZFLUX 1994 (Weddell Sea, winter), and SHEBA (1997–1998). Dot-dash curve is prediction according to Yaglom and Kader (1974) theory for heat and mass transfer over hydraulically rough surfaces

is required in interpreting the measurements. Nevertheless we are able to examine a number of different stations with significantly different underice z_0 values. Averages from several different experiments are shown in Fig. 6.5. Of the five station averages shown, the most complete data set by far is from SHEBA (McPhee 2002; McPhee et al. 2003), with an average value: $St_* = 0.0057 \pm 0.0004$. The average of all five stations indicated by the dashed line is 0.0056. An obvious inference from Fig. 6.5 is that St_* shows no discernible dependence on Reynolds number, when the latter is defined in terms of the friction velocity and roughness length, in apparent contradiction to the laboratory results. It is perhaps worth noting that in the laminar (Blasius) solution leading to (6.7), both the Stanton and Reynolds numbers are based on the "far-field" velocity, V, not u_{*0} or some other turbulent scale velocity (Incropera and DeWitt 1985). Regardless, it appears that with our definition of bulk Stanton number (6.10) in terms of u_{*0} and ΔT, it remains relatively constant over a wide range of conditions. This provides a critical constraint on values for the exchange coefficients α_h and α_S. Ignoring Reynolds number dependence, $\alpha_{h(S)} \propto (\nu/\nu_{T(S)})^{-n}$ and $R = \alpha_h/\alpha_S \approx (\nu_T/\nu_S)^n$.

For cold seawater, representative values for molecular diffusivities are $v_T = 1.39 \times 10^{-7}\,\mathrm{m^2\,s^{-1}}$ and $v_S = 6.8 \times 10^{-10}\,\mathrm{m^2\,s^{-1}}$ (Notz et al. 2003). From the laboratory results of Owens and Thomson (1963) and Yaglom and Kader (1974), n ranges from 2/3 to 0.8, implying $35 \leq R \leq 70$.

6.5 Double Diffusion—Melting

When sea ice melts there is strong evidence both by analogy with the laboratory studies and from the formation and vertical migration of "false bottoms" during the summer season (Notz et al. 2003, discussed below) that double diffusion is important. In principle, measurements during rapid melting of turbulent fluxes of momentum, heat, and salt near the interface, along with far-field temperature and salinity, would suffice to determine α_h and α_S, provided the interface is at freezing. In practice it is difficult to achieve unequivocal values, because (i) precise measurement of salinity flux is difficult, and (ii) rapid melting of sea ice, particularly pack ice in the marginal ice zone, often occurs in markedly heterogeneous environments with rapid temporal changes and large horizontal gradients.

The nonlinear character of (6.9) precludes a one-to-one relation between α_h and St_*; however, for a commonly observed range of forcing parameters, say $5 \leq u_{*0} \leq 15\,\mathrm{mm\,s^{-1}}$ and $0.05 \leq T_w \leq 0.5\,\mathrm{K}$ (and ignoring heat conduction or percolation in the ice column), the equation may be solved for an envelope of values for α_h over the range of R identified above that produce the same basal heat flux as the bulk formula (6.10) with $St_* = 0.0057$ (Fig. 6.6). Over most of the range of R, α_h is more than twice as large as St_*: when ice melts, interface salinity will be less than S_w, so $\delta T < \Delta T$ and $\alpha_h/St_* = \Delta T/\delta T$.

It is germane to ask whether the distinction between α_h and St_* is important. If the bulk coefficient is relatively constant as indicated by Fig. 6.5, is that not sufficient to gauge the basal heat flux? In most situations (i.e., conditions similar to

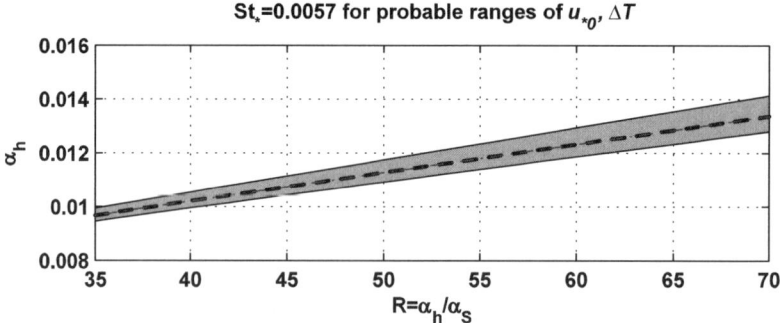

Fig. 6.6 Envelope of values of α_h that produce the same basal heat flux as the bulk formula $\langle w'T'\rangle_0 = St_* u_{*0} \Delta T$ for ranges $5 \leq u_{*0} \leq 15\,\mathrm{mm\,s^{-1}}$, $0.05 \leq \Delta T \leq 0.5\,\mathrm{K}$ when $St_* = 0.0057$

the relatively mild thermal forcing typical of most of the data used to derive the averages in Fig. 6.5), this is probably acceptable. However, it is often more extreme situations that are interesting: e.g., survival of a large area of ice blown over water that is relatively warm. The following example illustrates the impact of double-diffusive strength. Suppose that ice with moderate ocean stress ($u_{*0} = 0.015\,\text{m s}^{-1}$) drifts over water with $\Delta T = 0.1$ K and that $\dot{q}$ and w_p are negligible, more or less typical of summer in the central Arctic. Now, if $R = 1$ (no double diffusion) the solution for $\alpha_h = \alpha_S = 0.0058$ yields the same basal heat flux, $34\,\text{W m}^{-2}$, as the bulk relation for $St_* = 0.0057$ (the mean value for the SHEBA project). The interface temperature and salinity are only slightly different from the mixed layer values: $\delta T = 0.098$ K. If instead, double diffusion is strong, $R = 70$, then the solution which provides $34\,\text{W m}^{-2}$ requires $\alpha_h = 70\alpha_S = 0.0144$. In this case $\delta T = 0.040$ K. The melt rate is slightly over a centimeter per day.

Next consider a range in parameter space where measurements are scarce, but which is likely to be encountered often when ice drifts into open water. We leave u_{*0} unaltered but increase ΔT twenty fold to 2 K (water temperature slightly above 0°C), and consider the conditions at the interface for the two sets of α_h and α_S determined above. With $R = 1$, $\alpha_h = 0.0058$, the calculated heat flux is $681\,\text{W m}^{-2}$, so in this case the solution is almost exactly linear in ΔT, hence nearly the same as the bulk relation. In contrast, the calculated heat flux when $R = 70$ and $\alpha_h = 0.0144$ is $988\,\text{W m}^{-2}$, 45% greater. Approximate melt rates for the low and high double diffusive strengths are 22 and 32 centimeters per day, respectively. If one pictures a typical marginal ice zone where wind often pushes pack ice back and forth across an ocean temperature front, this simple thought experiment does indeed demonstrate that survivability of sea ice may be quite sensitive to the strength of double diffusion.

6.6 Double Diffusion and False Bottoms

Evidence for the importance of double diffusion during melting comes perhaps unexpectedly from a curiosity of the summer ice pack: false bottoms. These occur when concavities in the ice underside fill with fresh meltwater, which being in contact with seawater well below 0°C, forms a thin layer of ice at the freshwater/seawater interface. The phenomenon was documented nicely during the summer of 1975, when Arne Hanson maintained an array of depth gauges at the main AIDJEX station Big Bear near the center of the Beaufort Gyre in the Canadian Basin. Figure 6.7 (adapted from Notz et al. 2003) shows time series of bottom elevation (with respect to the upper surface) at seven of Hanson's thickness gauges, three initially deployed in thick ice (BB 4–6) and four others deployed in relatively thinner ice (BB 2,3,7,8). The thicker sites all show steady ablation through the summer, but at the thinner sites there is a significant increase in distance to the ice bottom starting at around day 200 (19 July), and in fact some of the thinner sites showed a net increase in thickness over the summer. Hanson attributed the increased thickness to formation of false bottoms. Notz et al. (2003) investigated the evolution of false

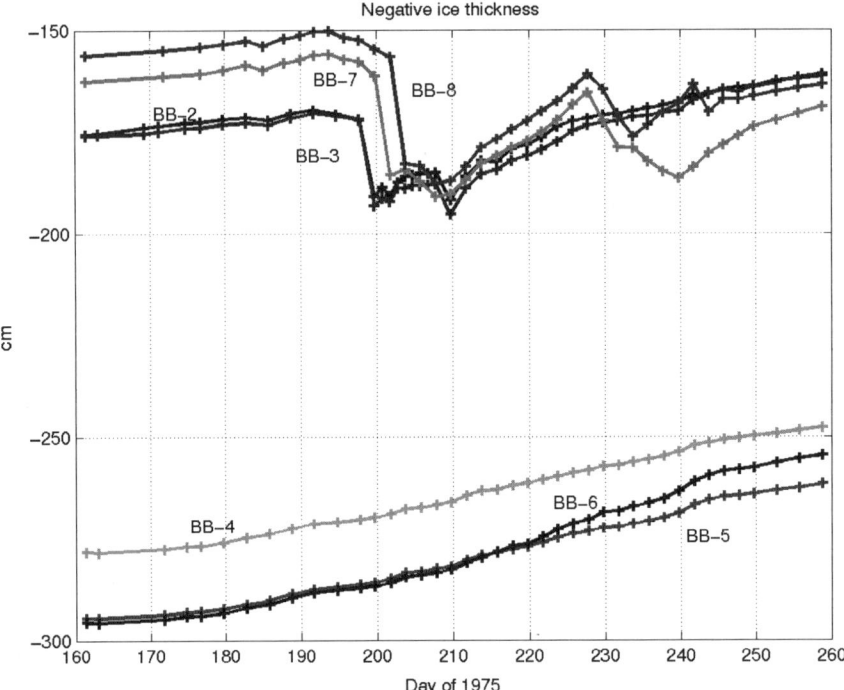

Fig. 6.7 Ice bottom elevation relative to the upper surface from ablation measurements made by A. Hanson during the 1975 AIDJEX project in the western Arctic (Adapted from Notz et al. 2003. With permission American Geophysical Union) (see also Plate 17 in the Colour Plate Section)

bottoms, in both laboratory and natural settings, including a simulation of the AIDJEX observations. The study showed that double diffusion is a critical process in the formation of false bottoms, which in turn may play an important role in maintenance of perennial sea ice.

In an idealized view of the false bottom layer after it has attained an initial finite thickness (Fig. 6.8), we assume horizontal homogeneity, and that the layer of water between the existing multiyear ice and the newly formed false bottom is fresh, with temperature equal to 0°C. We also assume that the thin ice layer is fresh with a linear temperature gradient. The false bottom thus borders on two different water types, and because it sustains a significant positive temperature gradient, there will be downward heat transfer. With these assumptions

$$\dot{q} = -\frac{K_{ice}(T_{up} - T_0)}{\rho_w c_p h} = -\frac{K_{ice} m S_0}{\rho_w c_p h} \quad (6.11)$$

where h is the thickness of the false bottom layer. The upper surface will migrate upward into the fresh water layer at rate

$$\dot{h}_{up} = -\frac{\rho_w \dot{q}}{\rho_{ice} Q_{fresh}}$$

6.6 Double Diffusion and False Bottoms

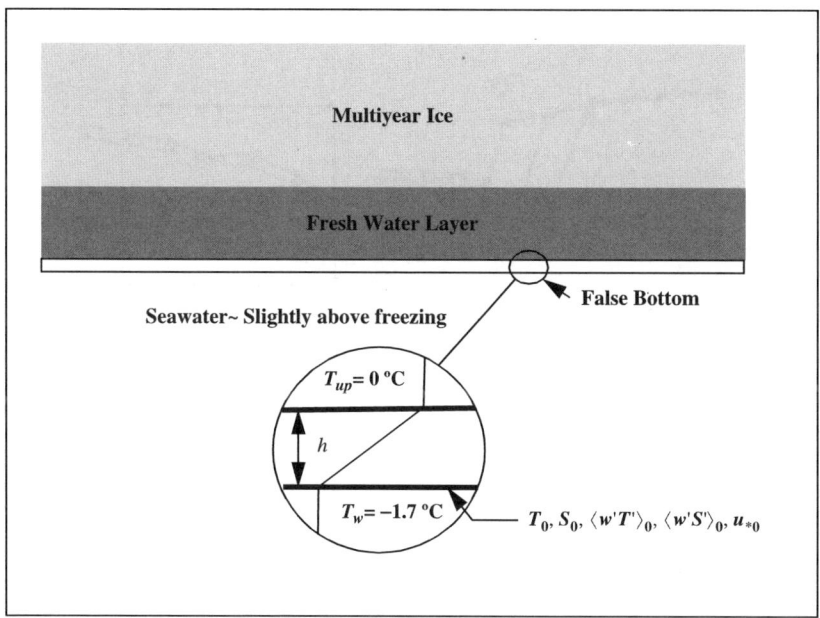

Fig. 6.8 Schematic of a thin layer of fresh ice that forms between meltwater trapped in underice concavities and colder seawater

where the latent heat of ice formation balances the downward heat flux. The lower boundary will also move vertically in response to the combination of $\dot{q}$ and $\langle w'T'\rangle_0$. The total rate of change of thickness for the false bottom layer is then

$$\dot{h} = \dot{h}_{up} - (\rho_w/\rho_{ice})w_0 = -(\rho_w/\rho_{ice})(\dot{q}/Q_{fresh} + w_0) \tag{6.12}$$

The bottom heat and mass balance can again be reduced to a quadratic equation for salinity at the interface

$$(1+\gamma_1)mS_0^2 + (T_w + T_L - (1+\gamma_1)mS_{ice})S_0 - T_wS_{ice} - T_LS_w = 0 \tag{6.13}$$

where $\gamma_1 = \frac{K_{ice}}{\rho_w c_p \alpha_h u_{*0} h}$.

Bottom ice elevation relative to a starting point on 9 July 1975 (year day 190) indicates false bottoms at four of Hanson's ice thickness sites (Fig. 6.9). They formed at different times, and there is some indication of multiple layer formation during a relatively calm period from day 205 to 210. But beginning with a period of more rapid ice drift from day 210 to 220, thickness decreased more or less uniformly at all four sites. Following the modeling approach of Notz et al. (2003), the evolution of a false bottom starting from an initial thickness of 2.5 cm was simulated by solving the combined equations (6.11), (6.12), and (6.13), with u_{*0} estimated from Rossby similarity (4.19) applied to the ice drift speed and with seawater temperature and salinity interpolated from daily CTD observations. We further specified that the

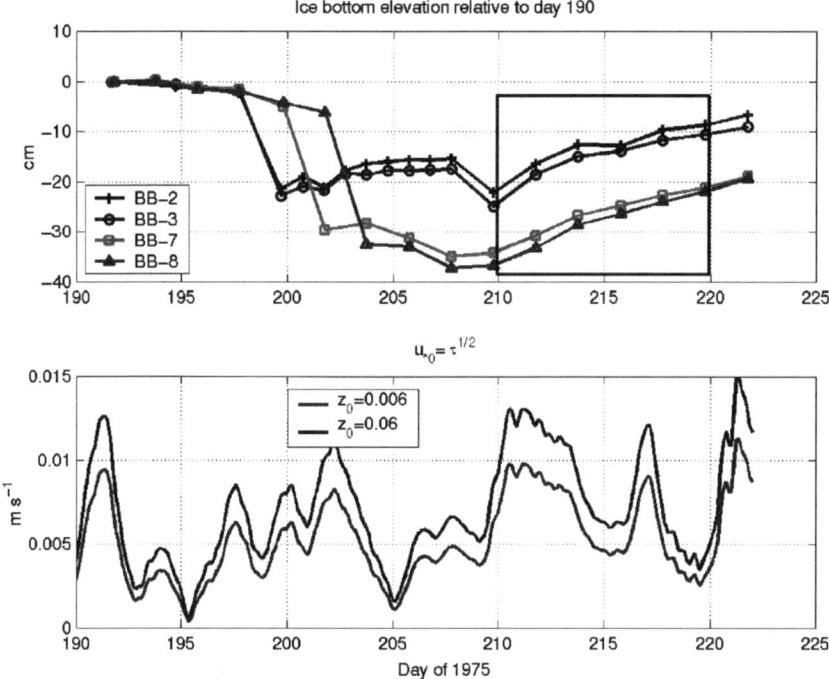

Fig. 6.9 Upper panel: bottom elevation of "false bottom" thickness gauges relative to their readings on day 190. Box marks the ten-day period chosen for simulation. Bottom panel: Interface friction velocity determined from ice drift relative to geostrophic current, for two values of surface roughness spanning range of estimates for AIDJEX station Big Bear (Adapted from Notz et al. 2003. With permission American Geophysical Union) (see also Plate 18 in the Colour Plate Section)

bulk Stanton number (6.10) be 0.0057, the mean SHEBA value, and $z_0 = 6$ mm, assuming the area around the false bottoms would be similar to the undeformed multiyear ice observed during SHEBA (McPhee 2002).

Model results for thick ice show about 6 cm of bottom ablation over the ten days (Fig. 6.10a), compared with about 14 cm of upward migration of the modeled false bottom (Fig. 6.10b). In each case the model matches Hanson's observations pretty well. The combination of $\alpha_h = 0.0111$ and $\alpha_h/\alpha_S = 50$ (which provides $St_* = 0.0057$, see Fig. 6.6) was chosen as the combination that minimized the root-mean square error between the model and observations in Fig. 6.10b, for R in the range 35 to 70. If the model is run with $R = 1$, with $\alpha_h = 0.0058$ (to maintain $St_* = 0.0057$) the results are reasonable for thick ice (Fig. 6.11a) but nonsensical for false bottom migration (Fig. 6.11b). The persistence of false bottoms in the summer pack is thus difficult to explain without invoking fairly strong double diffusion.

In Fig. 6.12a, modeled upward heat exchange between the ocean and thick ice is compared with the downward heat flux from false bottoms for the double-diffusive regime of Fig. 6.10. Because of the relatively large positive temperature gradient

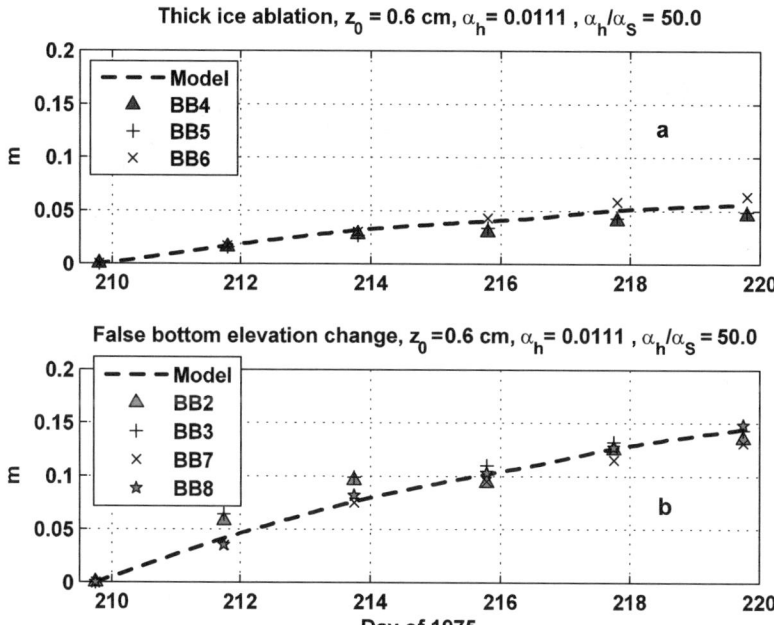

Fig. 6.10 a Bottom ablation under thick ice (gauges BB4–6) compared with model (dashed curves). **b** Elevation change at false bottom sites compared with model. Model parameters (listed) are the same (Adapted from Notz et al. 2003. With permission American Geophysical Union)

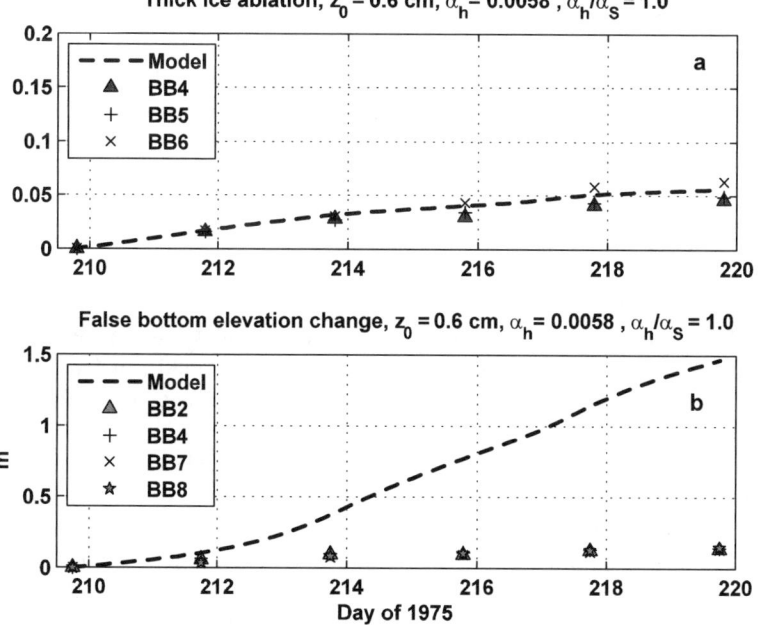

Fig. 6.11 As in Fig. 6.10, but for equal heat and salt exchange coefficients, chosen to maintain a realistic Stanton number

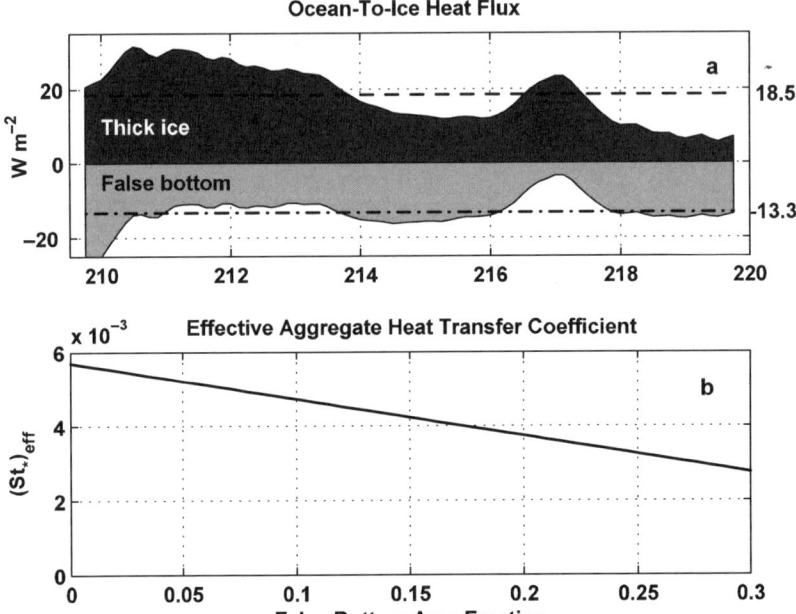

Fig. 6.12 a Time series of heat flux to thick ice (darker shading) and heat flux into the ocean from false bottoms (lighter). Average values are shown at right. **b** Aggregate Stanton number as a function of areal coverage of false bottoms and fresh water (Adapted from Notz et al. 2003. With permission American Geophysical Union) (see also Plate 19 in the Colour Plate Section)

across the false bottoms, they represent a source of heat for the upper ocean in summer apart from absorbed solar radiation penetrating into the well mixed layer. Depending on how ubiquitous false bottoms (or any freshwater/seawater interface on the ice undersurface) are during the melt season, they may exert a powerful influence on the total heat exchange between the IOBL and the pack ice. An effective aggregate Stanton number

$$(St_*)_{\text{eff}} = \frac{H_{\text{total}}}{\rho c_p u_{*0} \Delta T}$$

where $H_{\text{total}}/\rho c_p = (1 - A_{fb})\langle w'T'\rangle_0 + A_{fb}\langle w'T'\rangle_{fb}$ and A_{fb} is the area fraction of the undersurface covered by false bottoms or meltwater ponds, includes the combined positive and negative fluxes to thick ice ($\langle w'T'\rangle_0$) and from false bottoms ($\langle w'T'\rangle_{fb}$). It falls rapidly with increasing false bottom area. For the AIDJEX simulation, $(St_*)_{\text{eff}}$ is nearly halved if the area fraction approaches 3/10.

False bottoms and other manifestations of underice melt water may have a significant impact on the mass balance, and even the force balance, of the Arctic pack. Notz et al. (2003) reported estimates of false-bottom area fractions ranging from at least 10% (Jeffries et al. 1995) to over 50% (Hanson 1965). Jeffries et al. suggest that the origin of platelets in the Arctic ice cores they analyzed derived mainly from false bottom formation or an "ice-pump" mechanism, and that underice melt ponds may be more common than had been previously appreciated. Our experience while deploying the SHEBA station in the Beaufort Gyre in September 1997, was that when we drilled the late summer ice, we often encountered multiple layers of liquid meltwater interspersed between thin ice layers, suggestive of several successive cycles of false-bottom formation and migration. There was also a significant difference between establishing hydroholes during fall versus early spring. In the latter, it is often possible to extract ice that is dry to within 10–15 cm of the ice bottom, while in the former, we encountered a "water table" relatively high in the ice column below which lateral water movement appeared to be relatively unrestricted (and made further ice excavation more difficult). The concept of a porous water table that migrates downward from the surface as the ice column warms from above implies that any pre-existing concavities in the ice undersurface will be filled with fresh water regardless of a direct vertical connection to the surface.

False bottoms affect the general ice-albedo feedback issue in two important ways. First, they may substantially delay the transfer of heat from the upper ocean to the ice pack by reducing $(St_*)_{\text{eff}}$, which allows the upper ocean to maintain its heat content well past the time when sun angles are high. Second, as fresh water begins to collect in underice concavities early in the melt season, false bottom formation protects the thinnest ice from contact with the warming upper ocean thus delaying exposure of open seawater. The ice-albedo feedback is most effective when the ice/upper ocean system can absorb solar radiation at times near the summer solstice. Both of the false-bottom mechanisms described here tend to retard this timing, hence represent a perhaps important negative feedback in the system. A general thinning of the perennial pack (Rothrock et al. 1999) will mean that summer warming and the presumptive downward migrating "water table" will reach the ice base earlier in the summer, hence reinforcing the mitigating impact of underice melt ponds and false bottoms.

6.7 Freezing—Is Double Diffusion Important?

Mellor et al. (1986) and Steele et al. (1989) showed that if double-diffusive tendencies carry over to freezing in the same way that they apparently affect melting, then there ought to be significant production of supercooled water, because heat would be extracted from the upper ocean faster than salt would be injected. Presumably, the supercooled water would either nucleate in situ and form frazil ice crystals distributed in some way through the IOBL, or would nucleate more or less uniformly on the ice undersurface, regardless of ice thickness. Steele et al. (1989), using exchange parameters inferred from MIZEX measurements, estimated that supercooling and subsequent frazil production could account for as much as half of the ice accretion for thin (20 cm) ice to 30% for 80–100 cm thick ice.

If only one thickness of ice is considered, over time it matters little whether the ice forms from congelation at the immediate interface, or by accretion of frazil crystals drifting up to the interface from below. However, with different ice thicknesses there is a potentially interesting wrinkle. If a significant fraction of total ice production is in the form of frazil crystals, growth of categories near the low end of the thickness distribution will be slower than otherwise, while thick ice will accrete faster. If the growth of thin ice is retarded, its steep temperature gradient (responsible for most of the total heat transfer) will persist for a longer time, with the possibility of more overall heat transfer out of the ocean. Holland et al. (1997) examined this in a modeling study coupling an upper ocean model to an ice model with eight thickness categories. Using the "three-equation" parameterization suggested by McPhee et al. (1987), they found that the equilibrium annual average ice thickness increased by about 10 cm compared with an identical model run that was the same except that the exchange coefficients remained equal. There were substantial differences in modeled basal accretion.

In the multiyear ice pack of the Arctic, observations indicate that neither supercooling nor frazil production is extensive during winter. By examining thin sections in sea ice, it is relatively straightforward to distinguish between columnar ice accreted by congelation with horizontal c-axis orientation versus that from frazil, with more random orientation. Weeks and Ackley (1986) report that frazil accounts for only about 5% of total ice volume in Arctic pack ice and in fast sea ice from both hemispheres. It is found mainly near the surface, produced during initial ice formation. In the Antarctic, frazil-dominated structure is much more common, probably as a result of intense air-sea interaction in the vast marginal ice zones of the Southern Ocean. Over most of the Weddell Gyre, for example, the seasonal ice remains quite thin, often with a bi-modal thickness distribution from rafting by waves. Such conditions are conducive to frazil production.

While not common, supercooled water has been observed beneath the Arctic ice pack. Untersteiner and Sommerfeld (1964) reported supercooling of approximately 4 mK (i.e., water temperature about 0.004 K below its freezing temperature, dependent on salinity and pressure) near ice island ARLIS 2 (a drifting tabular berg) from measurements in water under the adjacent pack ice. They used a differential temperature measurement technique that did not require accurate salinity determination, an important consideration at the time. In that case, the supercooling was possibly attributable to the "ice-pump" effect described by Foldvik and Kvinge (1974) and Lewis and Perkin (1983). In typical pack ice, water in the well mixed IOBL will contact ice at varying pressures, e.g., ridge keels at pressures up to 10 dbar and beyond. Water that is at it freezing temperature at the level of the undeformed ice (say, 2 dbar) would be about 6 mK above freezing at 10 dbar.[3] The ice pump occurs when this water melts ice at depth, thus attaining a freezing temperature associated with the pressure where melting occurred. As this water rises following the ice morphology, it will be supercooled relative to its *in situ* pressure and will deposit ice as it encounters nucleation sites at the ice/water interface. In this way, ice can be transported through the thickness distribution from thicker to thinner categories. The ice pump is especially effective under floating ice shelves where large basal melting near the grounding line is "redeposited" as sea ice at higher levels near the terminus.

---

[3] We use the UNESCO formula for the freezing point of seawater from Millero (1978) as reported by Gill (1982), who points out that the formula fits measurements to an accuracy of ±4 mK.

6.7 Freezing—Is Double Diffusion Important?

Another possible source of supercooling can arise from differential mixing of salt and heat when there are large horizontal gradients in temperature and salinity. Measurements in 2007 of transient supercooling events in an energetic tidal flow in Fremansundet, Svalbard, were suggestive of this mechanism. The events occurred at different times at two different levels when a front between slightly less saline water from outside the fast ice was advected into the sound by the tide. The water on both sides of the front was within millikelvins of freezing, so that the water advected into the sound was slightly warmer. We interpreted the transient events (each lasting about an hour) as the result of heat mixing more rapidly than salt, so that the incoming water mass dipped below its freezing point in the frontal zone. It is perhaps worth noting that the supercooling north of Svalbard reported by Lewis and Perkin (1983) occurred in a region with strong horizontal gradients in temperature and salinity.

Given several possible sources of supercooling and subsequent frazil production (but a lack of evidence that it occurs extensively under multiyear pack ice), is it possible to examine in isolation the hypothesized mechanism of supercooling associated with double diffusion at the interface during rapid ice growth? We approached this problem as follows. Consider growth in thin ice in seawater at freezing under with the following conditions: $u_{*0} = 5$ mm s^{-1}, $S_w = 34$ psu, $T_w = T_f(S_w) = -1.865°$C, with upward heat conduction of 20 W m^{-2} in the ice column, corresponding to a temperature gradient in the ice of about -10 K m^{-1}. We assume $S_{ice} = 7$ psu. First solve the interface equation for salinity (6.9), with no double diffusion, i.e., $\alpha_h = \alpha_S = 0.0058$ (which was shown above to match the Stanton number constraint, $St_* = 0.0057$). In this case, $S_0 = 34.067$ psu, the ice grows at a rate of about 7 mm per day, and under the assumption that ice salinity is 7 psu, this produces a salinity flux $\langle w'S' \rangle_0 = -1.96 \times 10^{-6}$ psu m s^{-1}, and an upward heat flux from the water column of 0.4 W m^{-2}, which would be difficult to detect by covariance measurement. It is easy to confirm that $\langle w'T' \rangle_0 = -m \langle w'S' \rangle_0$, i.e., that heat is extracted from the water at just the rate required to maintain the water at its freezing temperature as salt is added at the surface. Note that none of these quantities are extreme by any measure.

Next solve the problem with identical conditions except that we now let double diffusion operate in the interface control volume at levels used in the false bottom simulation ($\alpha_h = 0.0111$; $\alpha_S = \alpha_h/50$, see Fig. 6.10). In this case $S_0 = 34.859$, congelation growth is significantly reduced to about 3.3 mm per day, and now the heat flux out of the water column is 10.7 W m^{-2}. This is easily measured, and thought experiments like this convinced us that the best way to look for the supercooling effect due to double diffusion at the freezing interface was by measuring upward heat flux in water just below the interface.

From these considerations, we designed a field experiment in a relatively controlled environment offered by fast ice and a gentle tidal flow in Van Mijen Fjord, Svalbard. In March 2001, we occupied a site on smooth fast ice, and installed instrumentation to measure ice characteristics and turbulence 1 m below the ice/water interface (McPhee et al. 2008, in press). Temperature profiles measured during the field project (Fig. 6.13) show the impact of changing surface temperature (there was

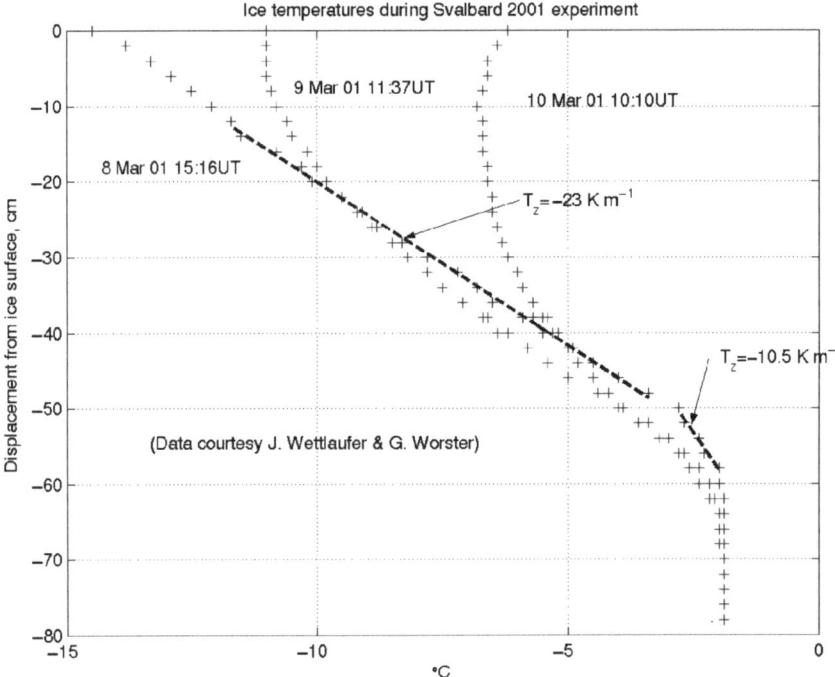

Fig. 6.13 Ice temperature profiles on three days in fast ice on VanMijen Fjord, Svalbard (see also Plate 20 in the Colour Plate Section)

little snow), but show a gradient in the lower 10 cm or so of the ice column that would indicate upward heat conduction there of about 21 W m^{-2}. Turbulence measurements are summarized in Fig. 6.14, where the turbulence data have been bin averaged according to mean current velocity 1 m below the interface in each 15-min turbulence realizations. A least-squares regression through the origin is quite close to the law of the wall for a hydraulically smooth boundary (Hinze 1975).

$$\frac{u_{1m}}{u_{*0}} = \frac{1}{\kappa} \log \frac{u_{*0}}{\nu} + 4.9$$

The regression lines in Fig. 6.14a and b show a slight correlation between flux magnitude and current speed, but are barely distinguishable from zero at the 90% confidence level. Overall the conditions are not much different from the example presented above, and the measured heat flux is only slightly more than what would be required to keep the well mixed layer near freezing as it became saltier. The lack of much ocean heat flux in this controlled environment is probably the most convincing evidence that during freezing double diffusion is relatively unimportant, in contrast to melting. By applying numerical model using local turbulence closure (described in Chapter 8) to longer term measurements from the VMF experiment, we showed that the exchange coefficient had to be close to unity during the VMF 2001 exercise (McPhee et al. 2008, in press).

6.7 Freezing—Is Double Diffusion Important?

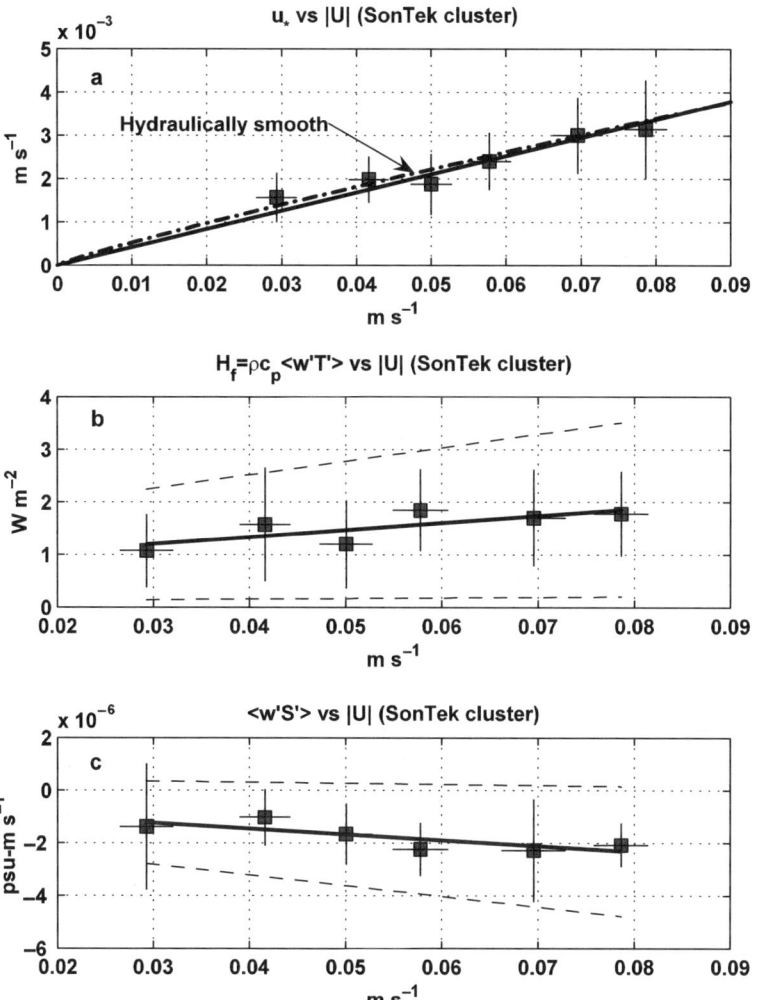

Fig. 6.14 Average turbulence measurements during the VMF exercise, binned according to tidal flow velocity. Error bars show ± one standard deviation in the 15-min turbulence realizations for each bin. (**a**) Friction velocity. Solid line is a least-squares fit through the origin; the dot-dashed curve is the law of the wall for a hydraulically smooth surface. (**b**) Turbulent heat flux. The solid line is a least squares regression for heat flux against current speed; the light dashed lines are confidence limits for the fit. (**c**) Same as b except for salinity flux

Why the freezing process should be fundamentally different from melting (in terms of double-diffusive effects in the IOBL) is apparently due to the fact that solidification occurs by the advance of mushy layers in which convection within the ice lattice relieves the double diffusive tendency: see, e.g., Wettlaufer et al. (1997), Notz (2005), and Feltham et al. (2006). During melting, the ice undersurface is observed to be uniformly smooth and the diffusion of salt into the crystal lattice apparently takes on a completely different character.

References

Barenblatt, G. I.: Scaling, Self-Similarity, and Intermediate Asymptotics. Cambridge University Press, Cambridge (1996)

Feltham, D. L., Untersteiner, N., Wettlaufer, J. S., and Worster, M. G.: Sea ice is a mushy layer. Geophys. Res. Lett., 33, L14501 (2006), doi: 10.1029/2006 GL026290

Foldvik, A. and Kvinge, T.: Conditional instability of sea water at the freezing point. Deep-Sea Res., 21, 160–174 (1974)

Gill, A. E.: Atmosphere-Ocean Dynamics. Academic, New York (1982)

Hanson, A. M.: Studies of the mass budget of arctic pack ice floes. J. Glaciol., 41, 701–709 (1965)

Hinze, J. O.: Turbulence, Second Edition. McGraw-Hill, New York (1975)

Holland, M. M., Curry, J. A., and Schramm, J. L.: Modeling the thermodynamics of a sea ice thickness distribution 2. Sea ice/ocean interactions. J. Geophys. Res., 102, 23,093–23,107 (1997)

Ikeda, M.: A mixed layer beneath melting sea ice in the marginal ice zone using a one-dimensional turbulent closure model, J. Geophys. Res., 91, 5054–5060 (1986).

Incropera, F. P. and DeWitt, D. P.: Fundamentals of Heat and Mass Transfer, Second Edition. Wiley, New York (1985)

Jeffries, M. O., Schwartz, K., Morris, K., Veazey, A. D., Krouse, H. R., and Cushing, S.: Evidence for platelet ice accretion in Arctic sea ice development. J. Geophys. Res., 100(C6), 10905–10914 (1995)

Josberger, E. G.: Sea ice melting in the marginal ice zone, J. Geophys. Res., 88, 2841–2844 (1983)

Lewis, E. L. and Perkin, R. G.: Supercooling and energy exchange near the Arctic Ocean surface. J. Geophys. Res., 88 (C12), 7681–7685 (1983)

Maykut, G. A. and Untersteiner, N.: Some results from a time-dependent thermodynamic model of sea ice. J. Geophys. Res., 76, 1550–1575 (1971)

Maykut, G. A.: An introduction to ice in polar oceans, Report APL-UW 8510, Applied Physics Laboratory, University of Washington, Seattle, WA (1985)

McPhee, M. G.: Turbulent stress at the ice/ocean interface and bottom surface hydraulic roughness during the SHEBA drift. J. Geophys. Res., 107 (C10) 8037 (2002), doi: 10.1029/2000JC000633

McPhee, M. G., Maykut, G. A., and Morison, J. H.: Dynamics and thermodynamics of the ice/upper ocean system in the marginal ice zone of the Greenland Sea. J. Geophys. Res., 92, 7017–7031 (1987)

McPhee, M. G., Kikuchi, T., Morison, J. H., and Stanton, T. P.: Ocean-to-ice heat flux at the North pole environmental observatory. Geophys. Res. Lett., 30 (24) 2274 (2003), doi: 10.1029/2003GL018580

McPhee, M. G., Morison, J. H., and Nilsen, F.: Revisiting heat and salt exchange at the ice-ocean interface: Ocean flux and modeling considerations, J. Geophys. Res., doi:10.1029/2007JC004383, in press (2008)

Mellor, G. L., McPhee, M. G., and Steele, M.: Ice-seawater turbulent boundary layer interaction with melting or freezing. J. Phys. Oceanogr., 16, 1829–1846 (1986)

Millero, F. J.: Freezing point of seawater, in: Eighth Report of the Joint Panel on Oceanographic Tables and Standards, UNESCO Tech. Pap. Mar. Sci. No. 28, Annex 6, UNESCO, Paris (1978)

Notz, D.: Thermodynamic and fluid-dynamical processes in sea ice. Ph. D. dissertation, Trinity College (2005)

Notz, D., McPhee, M. G., Worster, M. G., Maykut, G. A., Schlünzen, K. H., and Eicken, H.: Impact of underwater-ice evolution on Arctic summer sea ice. J. Geophys. Res., 108 (C7) 3223 (2003), doi:10.1029/2001JC001173

Owen, P. R. and Thomson, W. R.: Heat transfer across rough surfaces. J. Fluid Mech., 15, 321–334 (1963)

Parkinson, C. L. and Washington, W. M.: A large-scale numerical model of sea ice. J. Geophys. Res., 84, 311–337 (1979)

Rothrock, D. A., Yu, Y., and Maykut, G. A.: Thinning of the Arctic ice cover. Geophys. Res. Lett., 26, 3469–3472 (1999)

Steele, M., Mellor, G. L., and McPhee, M. G.: Role of the molecular sublayer in the melting or freezing of sea ice. J. Phys. Oceanogr., 19, 139–147 (1989)

Untersteiner, N. and Sommerfeld, R.: Supercooled water and bottom topography of floating ice. J. Geophys. Res., 69, 1057–1062 (1964)

Weeks, W. F. and Ackley, S. F.: The growth, structure, and properties of sea ice. In: N. Untersteiner (eds.) The Geophysics of Sea Ice, pp. 9–164. Plenum, New York (1986)

Wettlaufer, J. S., Worster, M. G., and Huppert, H. P.: Natural convection during solidification of an alloy from above with application to the evolution of sea ice, J. Fluid Mech., 344, 291–316 (1997)

Yaglom, A. M. and Kader, B. A.: Heat and mass transfer between a rough wall and turbulent flow at high Reynolds and Peclet numbers. J. Fluid Mech., 62, 601–623 (1974)

Nomenclature

$\dot{h}_i$	Ice growth rate
w_0	Isostatically adjusted bottom melt rate
w_p	Interface velocity due to freshwater percolation
w	Interface vertical velocity $(w_0 + w_p)$
$\dot{q}$	Heat conduction in the ice column divided by ρc_p
K_{ice}	Thermal conductivity of sea ice
K_{fresh}	Thermal conductivity of pure ice ($2.04\,\text{J m}^{-1}\,\text{K}^{-1}\,\text{s}^{-1}$)
S_{ice}	Ice salinity
T_z	Vertical temperature gradient in ice
β	Proportionality constant in Untersteiners (1961) formula ($0.117\,\text{J m}^{-1}\,\text{K}^{-1}\,\text{s}^{-1}\,\text{psu}^{-1}\,{}^\circ\text{C}$)
Q_L	Latent heat of melting for saline ice divided by c_p
L_{fresh}	Latent heat of pure ice ($335.5\,\text{kJ kg}^{-1}$)
α_h, α_S	Turbulent exchange coefficients for heat and salt
T_w, S_w	Far-field (well mixed layer) temperature, salinity
T_0, S_0	Temperature, salinity at the ice/water interface
$\delta T, \delta S$	$T_w - T_0$, $S_w - S_0$
Re_*	Reynolds number based on friction velocity, surface roughness: $u_{*0} z_0 / \nu$
Pr	Prandtl number, ν / ν_T
Sc	Schmidt number, ν / ν_S
St	Stanton number
St_*	Stanton no. based on u_{*0}; bulk heat transfer factor
ΔT	Departure of well mixed layer temperature from freezing

Chapter 7
A Numerical Model for the Ice/Ocean Boundary Layer

Abstract: A numerical model approach for simulating the IOBL is presented in this chapter. The staggered grid and implicit solution algorithms are patterned closely on techniques I learned while collaborating with George Mellor when we were both occupying visiting research chairs at the Naval Postgraduate School in Monterey, California. They were first used to model the IOBL with the Mellor-Yamada "level $2^1/_2$" second-moment closure (Mellor and Yamada 1982; Mellor et al. 1986) and later adapted to the first-order closure based on similarity scaling (McPhee et al. 1987). The latter is accomplished by expressing eddy viscosity and eddy diffusivity as the product of a local scale velocity and mixing length. It is essentially an implementation of the scaling principles described in Chapter 5, and will hereafter be referred to as *local turbulence closure* (LTC). LTC differs from the Mellor-Yamada and so-called $k - \varepsilon$ (e.g., Burchard and Baumert 1995) models in that the length scales are based on a combination of measurements and similarity theory, rather than derived from separate TKE and master length scale conservation equations. A practical impact is that the LTC model eliminates the need to carry these equations in the solution.

We start with a review of a fairly standard leap-frog-in-time, implicit solution technique on a staggered vertical grid, and explore various approaches to specifying boundary conditions. We then discuss the algorithms for calculating the mixing length and eddy viscosity under varying conditions of buoyancy flux in the IOBL, and match the fluid model to an algorithm implementing the interface conditions described in Chapter 6. The model is exercised for several examples in Chapter 8.

7.1 Difference Equations

The partial differential equation for a conserved quantity in a horizontally homogeneous fluid is

$$\theta_t = -F_z^\theta + Q^\theta \quad (7.1)$$

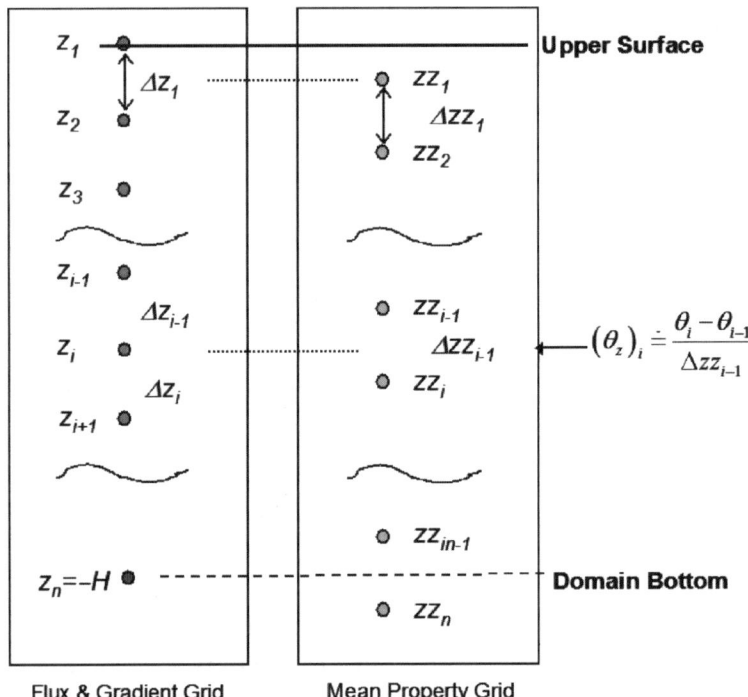

Fig. 7.1 Staggered grid scheme. Mean properties are evaluated on the zz grid, flux and gradient related quantities on the z grid

where subscripts denote partial differentiation, θ is a generic conservative property of the fluid, F^θ is the vertical flux of θ, and Q^θ is a source term. With the leap-frog-in-time scheme, the corresponding difference equation is

$$\frac{\theta_{i,j+1} - \theta_{i,j-1}}{2\Delta t} = -\frac{(F_{i+1}^\theta - F_i^\theta)}{\Delta z_i} + Q^\theta \tag{7.2}$$

where fluxes are evaluated on a staggered grid (Fig. 7.1). The grid is chosen so that a conserved quantity (where θ might represent temperature, salinity, or momentum) is evaluated at points on the zz grid, where zz_i lies midway between points z_i and z_{i+1} in the z grid, on which variables dependent on gradients (fluxes, buoyancy frequency, eddy viscosity, etc.) are evaluated.

First-order closure relates flux to the mean quantity vertical gradient by means of an eddy diffusivity appropriate to θ

$$F^\theta = -K^\theta \theta_z \tag{7.3}$$

with corresponding difference equation

$$F_i^\theta = \frac{-K_i^\theta}{\Delta zz_{i-1}} (\theta_i - \theta_{i-1}) \tag{7.4}$$

7.2 Boundary Conditions

Substituting (7.4) into (7.2) and grouping terms

$$A_i \theta_{i+1,j+1} + B_i \theta_{i,j+1} + C_i \theta_{i-1,j+1} = \theta_{i,j-1} + 2\Delta t Q_i^\theta \tag{7.5}$$

where

$$A_i = \frac{-2\Delta t K^\theta{}_{i+1}}{\Delta z_i \Delta z z_i} \quad C_i = \frac{-2\Delta t K^\theta{}_i}{\Delta z_i \Delta z z_{i-1}} \quad B_i = 1 - A_i - C_i$$

From here on, the second index referring to the new $(j+1)$ time step will be dropped from the notation, with only the previous $(j-1)$ time step indicated explicitly. To use the implicit solution technique, let

$$\theta_i = D_i + E_i \theta_{i+1} \tag{7.6}$$

so that

$$A_i \theta_{i+1} + B_i \theta_i + C_i (D_{i-1} + E_{i-1} \theta_i) = \theta_{i,j-1} + 2\Delta t Q_i^\theta$$

from which

$$\theta_i = \frac{-A_i \theta_{i+1} + \theta_{i,j-1} + 2\Delta t Q^\theta{}_i - C_i D_{i-1}}{B_i + C_i E_{i-1}} \tag{7.7}$$

hence the recursion relation is

$$E_i = \frac{A_i}{A_i - 1 + C_i(1 - E_{i-1})} \tag{7.8}$$

$$D_i = \frac{C_i D_{i-1} - \left(\theta_{i,j-1} + 2\Delta t Q_i^\theta\right)}{A_i - 1 + C_i(1 - E_{i-1})}$$

This approach holds as well for the momentum conservation equation in a rotating reference frame provided the Coriolis term is treated as a source (time centered in the leapfrog scheme). Using complex notation for the horizontal velocity (relative to undisturbed geostrophic flow) in a horizontally homogeneous fluid, the momentum balance is

$$u_t = \tau_z - ifu \tag{7.9}$$

where τ_z is the vertical gradient of the kinematic Reynolds stress in the fluid. Note that an additional source term (e.g., a constant or depth varying ["thermal wind"] geostrophic current) may be added to the Coriolis source term.

7.2 Boundary Conditions

The starting point for the recursion relations (7.8) is provided by consideration of the boundary conditions for the difference form of the conservation equation for the generic property θ.

7.2.1 Flux of Variable θ Specified at Upper Surface

If the flux of θ is specified at the upper boundary, the difference equation for the gridpoint zz_1 is

$$\theta_1 - \theta_{1,j-1} = -F_2{}^\theta \frac{2\Delta t}{\Delta z_1} + F_1{}^\theta \frac{2\Delta t}{\Delta z_1} + 2\Delta t Q_1{}^\theta = \frac{2\Delta t K_2{}^\theta}{\Delta z_1 \Delta zz_1}(\theta_2 - \theta_1) + 2\Delta t \left(\frac{F_1{}^\theta}{\Delta z_1} + Q_1{}^\theta\right) \quad (7.10)$$

or grouping terms

$$(1 - A_1)\theta_1 = -A_1\theta_2 + 2\Delta t\left(\frac{F_1{}^\theta}{\Delta z_1} + Q_1{}^\theta\right) + \theta_{1,j-1} \quad (7.11)$$

so

$$E_1 = \frac{A_1}{A_1 - 1}$$

$$D_1 = \frac{-2\Delta t\left(\frac{F_1{}^\theta}{\Delta z_1} + Q_1{}^\theta\right) - \theta_{1,j-1}}{A_i - 1} \quad (7.12)$$

7.2.2 Variable θ Specified at Upper Surface

A second class of boundary condition addresses the case where the value of θ is specified at the interface instead of its flux. In this case, an estimate of the surface flux is made from the surface value, θ_s, and the value at the first grid point using the time-centered estimate of friction velocity

$$F_s{}^\theta = \frac{u_{*0}(\theta_1 - \theta_s)}{\Phi^\theta} \quad (7.13)$$

where Φ^θ is the dimensionless change in the mean quantity across the distance separating the surface and the first grid point

$$\Phi^\theta = u_{*0} \int_0^{|zz_1|} \frac{1}{\mathfrak{R}} dz \quad (7.14)$$

where $\mathfrak{R}$ is the effective viscosity or diffusivity (not necessarily the eddy diffusivity).

The difference equation for the first grid point is

$$\theta_{1,j+1} - \theta_{1,j-1} = -F_2{}^\theta \frac{2\Delta t}{\Delta z_1} + F_1{}^\theta \frac{2\Delta t}{\Delta z_1} + 2\Delta t Q_1{}^\theta \quad (7.15)$$

$$= -A_1(\theta_2 - \theta_1) + s_I(\theta_2 - \theta_1) + 2\Delta t Q_1{}^\theta$$

7.2 Boundary Conditions

where

$$s_l = \frac{2\Delta t u_{*0}}{\Phi^\theta \Delta z_1}$$

from which

$$E_1 = \frac{A_1}{A_1 + s_l - 1} \tag{7.16}$$

$$D_1 = \frac{s_l \theta_s - 2\Delta t Q_1^\theta - \theta_{1,j-1}}{A_i + s_l - 1}$$

For momentum, the dimensionless velocity change from the surface to the first grid point is given by the law of the wall, provided $|zz_1|$ is much less than the Obukhov length, i.e.,

$$\Phi^u = \frac{1}{\kappa} \ln \left| \frac{zz_1}{z_0} \right| \tag{7.17}$$

where z_0 is undersurface roughness length. If the grid is relatively coarse and surface buoyancy flux is significant, Φ^u may be estimated from Monin-Obukhov theory with some correction for stress rotation (McPhee 1990).

For the ice/ocean interface we have found that the dimensionless changes in temperature and salinity near the boundary are much greater than for momentum (by several orders of magnitude for salinity) complicated during melting by double-diffusive effects (Chapter 6). Consequently the approach taken is to invoke a sub-model for the heat and salt exchange at the interface, which provides flux boundary conditions for the heat and salt conservation equations (Section 7.7).

7.2.3 Dynamic Momentum Flux Condition

A special case exists for wind driven sea ice drift when internal ice stress gradients are small relative to other forces, but where the inertia of the solid ice cover is important in the overall momentum balance. The following is an approach formulated by G. Mellor (1984, personal communication). For ice draft ("equivalent ice thickness"), h_{ice}, the ice momentum equation is

$$h_{\text{ice}} \left(\frac{\partial \boldsymbol{u}_s}{\partial t} + if \boldsymbol{u}_s \right) + \boldsymbol{\tau}_w - \boldsymbol{\tau}_a = 0 \tag{7.18}$$

where τ_a is air stress at the ice upper surface divided by water density. To account for shear between the uppermost mean quantity gridpoint and the interface, surface velocity is expressed as

$$\boldsymbol{u}_s = \boldsymbol{u}_1 + \boldsymbol{u}_{*0} \Phi^u$$

where Φ^u is given by (7.17). The difference form of (7.18) is

$$2\Delta t \boldsymbol{\tau}_w = -h_{\text{ice}}(\boldsymbol{u}_{1j+1} - \boldsymbol{u}_{1j-1}) + 2\Delta t \boldsymbol{\tau}_m \tag{7.19}$$

$$\boldsymbol{\tau}_m = \boldsymbol{\tau}_a + h_{\text{ice}} \boldsymbol{Q}_1^u - h_{\text{ice}} if \Phi^u$$

which incorporates both the dynamic aspect of ice inertia and the wind stress modified by the ice Coriolis force. Substituting into the momentum difference equation for the first grid point:

$$u_{1,j+1} - u_{1,j-1} = -\tau_2 \frac{2\Delta t}{\Delta z_1} + \tau_1 \frac{2\Delta t}{\Delta z_1} + 2\Delta t Q^u{}_1 \qquad (7.20)$$

where $\tau_1 = -\tau_w$, providing a modified version of (7.12)

$$E_1 = \frac{A_1}{A_1 - 1 + \frac{h_{ice}}{\Delta z_1}} \qquad (7.21)$$

$$D_1 = \frac{u_{1,j-1}\left(1 - \frac{h_{ice}}{\Delta z_1}\right) + 2\Delta t \left(\frac{-\tau_m}{\Delta z_1} + Q_1{}^u\right) - \theta_{1,j-1}}{A_i - 1 - \frac{h_{ice}}{\Delta z_1}}$$

7.2.4 Flux of θ Specified at the Bottom of Model Domain

For a flux condition specified at the lower boundary (including zero flux)

$$\theta_{kb-1} - \theta_{kb-1,j-1} = \frac{2\Delta t}{\Delta z_{kb-1}} \left(F_b^\theta - F^\theta{}_{kb-1}\right) + 2\Delta t Q^\theta{}_{kb-1} \qquad (7.22)$$

substituting for F^θ_{kb-1} and rearranging terms

$$C_{kb-1}\theta_{kb-2} = (C_{kb-1} - 1)\theta_{kb-1} + 2\Delta t \left(Q^\theta{}_{kb-1} + \frac{F_b^\theta}{\Delta z_{kb-1}}\right) + \theta_{kb-1,j-1} \qquad (7.23)$$

but since $\theta_{kb-2} = D_{kb-2} + E_{kb-2}\theta_{kb-1}$ we have

$$\theta_{kb-1} = \frac{2\Delta t \left(Q^\theta{}_{kb-1} + \frac{F_b^\theta}{\Delta z_{kb-1}}\right) + \theta_{kb-1,j-1} - C_{kb-1}D_{kb-2}}{1 + C_{kb-1}(E_{kb-2} - 1)} \qquad (7.24)$$

which then begins the iteration back up through the IOBL as indicated by (7.6).

7.2.5 θ Specified at the Bottom of the Model Domain

Figure 7.1 illustrates that zz_{kb} is beyond the model depth domain. Eddy viscosity/diffusivity is evaluated on the z (flux) grid and its last value is K^θ_{kb}, so from (7.5), A and the fields that depend on it are evaluated at indices from 1 to kb-1. Thus to specify the value at the base of the domain, set $\theta_{kb} = \theta_b$, from which

$$\theta_{kb-1} = D_{kb-1} + E_{kb-1}\theta_b \qquad (7.25)$$

which again starts the iterative evaluation of (7.6).

7.3 Steady-State Momentum Equation

The Ekman equation admits a steady-state solution of the momentum conservation equation, often a valuable approximation for interpreting isolated measurements (Chapter 9) or even calculating time evolution of the scalar variables (McPhee 1999). The steady version of (7.9) may be written

$$if\mathbf{u} = (K\mathbf{u}_z)_z \tag{7.26}$$

Using the same staggered grid as before, the difference equation is

$$\tilde{A}_i \mathbf{u}_{i+1} + \tilde{\mathbf{B}}_i \mathbf{u}_i + \tilde{C}_i \mathbf{u}_{i-1} = 0 \tag{7.27}$$

where

$$\tilde{A}_i = \frac{K_{i+1}}{\Delta z_i \Delta z z_i} \quad \tilde{C}_i = \frac{K_i}{\Delta z_i \Delta z z_{i-1}} \quad \mathbf{B}_i = -(if + \tilde{A}_i + \tilde{C}_i) \tag{7.28}$$

with $\mathbf{u}_i = \tilde{\mathbf{D}}_i + \tilde{\mathbf{E}}_i \mathbf{u}_{i+1}$, the recursion relations are

$$\tilde{\mathbf{D}}_i = \frac{-\tilde{C}_i \tilde{\mathbf{D}}_{i-1}}{\tilde{\mathbf{B}}_i + \tilde{C}_i \mathbf{E}_{i-1}} \tag{7.29}$$

$$\tilde{\mathbf{E}}_i = \frac{-\tilde{A}_i}{\tilde{\mathbf{B}}_i + \tilde{C}_i \mathbf{E}_{i-1}}$$

The momentum equation at gridpoint zz_1 is

$$if\mathbf{u}_1 = \tilde{A}_1 (\mathbf{u}_2 - \mathbf{u}_1) + \frac{-\tau_0}{\Delta z_1} \tag{7.30}$$

where $\tau_0 = -(\langle u'w' \rangle_0 + i \langle u'v' \rangle_0)$.

Using flux (stress) boundary conditions, the recursion calculation is started with

$$\tilde{\mathbf{D}}_1 = \frac{-\tau_0}{\Delta z_1 (\tilde{A}_1 + if)} \tag{7.31}$$

$$\tilde{\mathbf{E}}_1 = \frac{\tilde{A}_1}{\tilde{A}_1 + if}$$

7.4 Distributed Sources

Solar heating is an example of a scalar source term for which the flux at the surface is not necessarily the proper boundary condition for the conservation equation. Let $Q^\theta(z)$ be the source function of an arbitrary variable θ, for which the total flux through the surface is F_0^θ. To conserve θ

$$F_0^\theta = -\int_{-\infty}^{0} Q^\theta(z) dz \tag{7.32}$$

For an exponentially decaying source like shortwave radiation

$$Q^\theta = -\frac{F_0^\theta}{l_\theta} e^{z/l_\theta}$$

where l_θ is the "e-folding" depth. The attenuation of short wave radiative flux is a function of the clarity of the water and the spectral characteristics of the radiation. If multiple wavelengths are considered, then surface flux may be partitioned into categories characterized by different scale depths, and summed.

Another distributed source of potential interest in the IOBL problem is frazil crystallization. If the frazil crystals are generated near the surface and mixed downward, the problem bears much similarity to sediment transport in bottom boundary layers, where a "settling" velocity is mimicked by the rise rate of the slightly buoyant crystals. If on the other hand, the crystals nucleate *in situ* from supercooled water, in addition to a distributed source for ice concentration in the fluid, they also contribute distributed source terms for both heat and salt.

7.5 Solution Technique

The basic solution technique is available with these equations. For each of the primary model variables (typically u, T, S), an $n \times 3$ matrix is carried by the solution algorithm where n is the number of vertical gridpoints, and the second index refers to the $j-1$, j, and $j+1$ time steps, respectively. The second column represents the state of the system after the last iteration, and may be used to calculate fluxes via the appropriate difference equations, generically described by (7.4). As described below, these fluxes will be used along with the boundary fluxes to determine the eddy viscosity/diffusivity profiles for the next time step. For each time step, the prescribed surface boundary conditions, using options described in Section 7.2, are used to calculate D_1 and E_1 for each of the primary variables. The recursion relations (7.8) are solved for the remainder of the D and E arrays. The bottommost value for each variable is calculated according to the specified bottom boundary condition, with progressively higher grid points evaluated using (7.6). The leapfrog scheme calculates the third column from the first via (7.2) using source terms like the Coriolis force in (7.9) from the second column. To complete the time step, the first column $(j-1)$ is replaced by a weighted average of all three columns, and the second (j) is replaced by the updated quantity $(j+1)$. The system is now ready for the next time step.

7.6 The Local Turbulence Closure Model

The scales identified in Chapter 5, as summarized in Fig. 5.17, form the basis for *local turbulence closure* (LTC). The essential idea is that the eddies responsible for

7.6 The Local Turbulence Closure Model

the bulk of the turbulent transfer, characterized by the length scale λ, grow with distance from the boundary until they reach a limiting size determined by the interplay between shear stress and buoyancy flux at the interface, which are characterized by fractions of the planetary scale and the Obukhov length. Eddies of this size then fill the remaining well mixed portion of the IOBL.[1] The eddy viscosity is determined by the product of mixing length and *local* turbulent scale velocity.

If the surface forcing is strong enough to induce mixing in the pycnocline underlying the well mixed layer, the exchanges are characterized by a secondary boundary layer in the pycnocline, with scales determined by the stress and buoyancy flux at the interface between the well-mixed layer and the pycnocline. Hence the model must keep track of the depth of the pycnocline, z_{pyc}, which is typically accomplished by testing the buoyancy frequency squared against a limiting value N^2_{min}, i.e., determining the depth at which $N^2 = (-g/\rho)\rho_z$ first exceeds N^2_{min}.

A flow chart for determining λ in the well mixed portion of the IOBL is presented in Fig. 7.2. On the stable (right hand) side, the calculation is straightforward, with

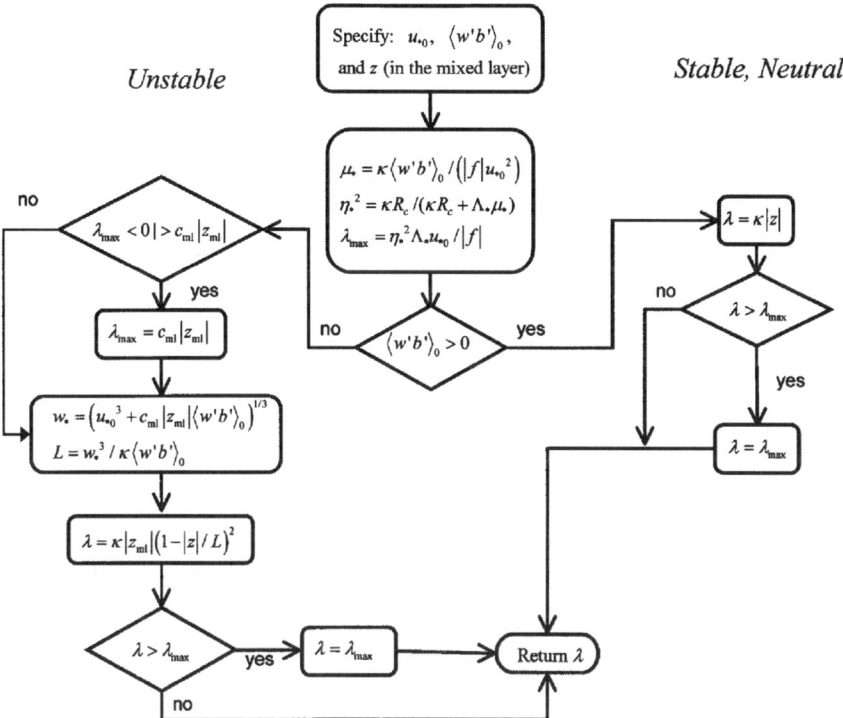

Fig. 7.2 Flow chart describing the algorithm for calculating mixing length in the well mixed portion of the IOBL

---

[1] By "well mixed" is meant where scalar property gradients are small, but not necessarily absent. When we measure scalar fluxes in the IOBL, we invariably find gradients, albeit tiny (see, for example, Figs. 5.4 and 5.15).

the stability factor η_*^2 reducing the size of the eddies (and the penetration of turbulent mixing) when interface melting is significant. In the surface layer, λ increases with distance until it reaches λ_{max}. For freezing (statically unstable) conditions, the situation is more complicated, because in a convective regime turbulence can mix with little or no shear. If buoyancy flux is strong enough that $\lambda_{max} = \eta_*^2 \Lambda_* u_{*0}/|f|$ is negative or larger than $c_{ml}|z_{pyc}|$, it is replaced with $\lambda_{max} = c_{ml}|z_{pyc}|$ where, based on LeadEx measurements, $0.2 < c_{ml} < 0.4$. In this way, for say a gradual transition to freezing, the model is capable of accepting mild convective conditions without a sudden shift in eddy size and eddy viscosity.[2]

The algorithm for the determining λ in the upper pycnocline is similar, with input u_{*p}, $\langle w'b' \rangle_p$, and $z - z_p$ substituted for the corresponding input values in Fig. 7.2. Generally, only the right-hand side (stable) would apply, and λ would normally be much smaller than in the well mixed layer. However, the model can readily handle a weak density gradient high in the IOBL (e.g., Fig. 5.17b), in which case u_{*p} and $\langle w'b' \rangle_p$ might be comparable to the interface values.

Once a distribution of mixing length is determined, the model calculates eddy viscosity as the product λ and the local turbulent scale velocity also determined from the previous time step. Most of the time, the latter is the square root of the local shear stress $(u_* = |K\mathbf{u}_z|^{1/2})$ unless there is negative buoyancy flux in the domain, in which case the scale velocity for those grid points is $w_* = (u_*^3 + c_{ml}d_{ml}\langle w'b' \rangle)^{1/3}$.

In statically unstable or near neutrally buoyant conditions, scalar eddy diffusivity is assumed to be the same as eddy viscosity (Reynold's analogy). In stably stratified flow, as encountered in the upper part of the pycnocline, momentum exchange, which depends on pressure fluctuations as well as direct mixing, is more efficient than scalar exchange. For lack of definitive geophysical measurements, the ratio of eddy diffusivity to eddy viscosity is specified by a formula that approximates laboratory results compiled by Turner (1973) relating the ratio of salt to momentum transfer coefficients. The *ad hoc* formula is

$$\frac{K_{h,S}}{K_m} = \begin{cases} 1 & Ri \leq 0.079 \\ \exp(-1.5\sqrt{Ri - 0.079}) & 0.079 < Ri \leq 5 \\ 0.039 & Ri > 5 \end{cases}$$

Where Ri is the gradient Richardson number.

The reduction of scalar eddy diffusivity relative to eddy viscosity in stratified flow begs the question of whether the haline and thermal diffusivities differ. Unfortunately, there is little hard evidence from natural boundary layer flows in salt water from which to draw conclusions regarding a ratio of, say, K_h/K_S. We found during the upwelling event described in Sections 2.6 and 2.7 (see Figs. 2.13 and 2.14), that heat was mixed out of the upper pycnocline more rapidly than salt, enough to significantly modify the T/S properties relative to the ambient surroundings (McPhee et al. 2005). Intuitively, we suspect that as stability increases and turbulence scales

---

[2] The mixing length scheme also allows the model to function near the equator, although this is moot for the IOBL.

diminish, the large disparity between molecular thermal and haline diffusivities will effect differential exchange, even if the flow remains sluggishly turbulent. The question calls out for more research.

7.7 The Ice/Ocean Interface Submodel

An interface submodel is used to implement the ice/ocean interface characteristics discussed in Chapter 6 as follows. The external forcing parameters $\dot{q}$ (kinematic conductive heat flux in the ice), S_{ice} (ice salinity), w_p (percolation velocity), and u_{*0} are assumed specified as driving time series, or perhaps provided by a separate ice model. Internal (to the model) variables are provided from the previous time step: T_w, S_w, d_{ml}, and $\langle w'b'\rangle_0$. If the last is zero or positive, the submodel assumes a melting or stationary interface, and uses specified double diffusive exchange coefficients. For example, $\alpha_h = 0.011$, $\alpha_S = \alpha_h/50$ would be reasonable choices. It sets the turbulent velocity scale to u_{*0}.

If on the other hand, the previous time step indicates freezing ($\langle w'b'\rangle_0 < 0$), the submodel sets $\alpha_h = \alpha_S$ (say, 0.0058), and modifies the turbulent scale velocity to be $w_{*0} = (u_{*0}^3 - c_{ml}d_{ml}\langle w'b'\rangle_0)^{1/3}$. The submodel then solves for S_0 as in (6.9), with fluxes given by (6.4), using in each case the appropriate velocity scale (u_{*0} or w_{*0}). The submodel then calculates the new value for buoyancy flux

$$\langle w'b'\rangle_0 = (g/\rho)(\beta_S\langle w'S'\rangle_0 - \beta_T\langle w'T'\rangle_0) \qquad (7.33)$$

where β_S and β_T are the haline contraction and thermal expansion factors evaluated at T_0 and S_0.

References

Burchard, H. and Baumert, H.: On the performance of a mixed layer model based on the turbulence closure. J. Geophys. Res., 100, 8523–8540 (1995)

McPhee, M. G.: A time-dependent model for turbulent transfer in a stratified oceanic boundary layer. J. Geophys. Res., 92 (C7), 6977–7986 (1987)

McPhee, M. G.: Small scale processes. In: Smith, W. (ed.) Polar Oceanography, pp. 287–334. Academic Press, San Diego, CA (1990)

McPhee, M. G.: Scales of turbulence and parameterization of mixing in the ocean boundary layer. J. Mar. Syst., 21, 55–65 (1999)

McPhee, M. G., Maykut, G. A., and Morison, J. H.: Dynamics and thermodynamics of the ice/upper ocean system in the marginal ice zone of the Greenland Sea. J. Geophys. Res., 92, 7017–7031 (1987)

McPhee, M. G., Kwok, R., Robins, R., and Coon, M.: Upwelling of Arctic pycnocline associated with shear motion of sea ice. Geophys. Res. Lett., 32, L10616 (2005), doi: 10.1029/2004GL021819

Mellor, G. L. and Yamada, T.: Development of a turbulence closure model for geophysical fluid problems. Rev. Geophys., 20, 851–875 (1982)

Mellor, G. L., McPhee, M. G., and Steele, M.: Ice-seawater turbulent boundary layer interaction with melting or freezing. J. Phys. Oceanogr., 16, 1829–1846 (1986)

Turner, J. S.: Buoyancy Effects in Fluids. Cambridge University Press, London (1973)

Nomenclature

θ	Generic conservative fluid property
θ_z	Partial derivative with respect to vertical coordinate
θ_t	Partial derivative with respect to time
F^θ, $F^\theta{}_z$	Flux of θ; partial derivative wrt z
Q^θ	Source term for property q
K^θ	Eddy diffusivity of property q
z_i	ith grid point on the property gradient grid
zz_i	ith grid point on the mean property grid
Φ^θ	Dimensionless change in θ from the boundary to zz_1
τ_a	Air stress on the upper ice surface divided by water density
τ_w	Water stress on the lower ice surface divided by water density
h_{ice}	Ice draft (thickness $\times \rho_{ice}/\rho_w$)
l_θ	E-folding length scale for exponentially decaying source
w_{*0}	Model scale velocity for statically unstable flows
β_T, β_S	Thermal expansion factor, saline contraction factor

Chapter 8
LTC Modeling Examples

Abstract: This chapter explores several features of the IOBL by combining observations and modeling based on *local turbulence closure* as incorporated into a numerical model described in Chapter 7. The intent is to elucidate certain features of the response of the upper ocean to variations in forcing that require consideration of the time dependence of the physical conservation equations.

First, we show that an interesting series of upper ocean measurements at the SHEBA site near the time of maximum insolation, when there was a clearly discernible diurnal signal in both temperature and downward turbulent heat flux at two measurement levels, can be adequately simulated. However, the simulation makes sense only if solar radiation penetrating the compact ice cover is significantly greater than has been typically assumed in the past.

Next is a simulation of events observed in late summer at the SHEBA site, when there was energetic inertial motion of the ice and upper ocean. Inertial oscillation nearly always implies strong shear in the upper part of the pycnocline, and early models of mixed-layer evolution (e.g., Pollard et al. 1973; Niiler and Kraus 1977) related the rate of mixed-layer deepening ("entrainment velocity") to a Richardson number involving the inverse square of the velocity of a uniform slab of water (volume transport divided by mixed-layer depth). In the slab model of Pollard et al. (1973), for example, the velocity was inertial and any deepening was confined to the first half inertial period unless the inertial velocity increased. This was an unrealistic limitation and much effort was devoted to elaborating how entrainment would take place at the base of the mixed layer, while still retaining the simplicity of constant temperature, salinity, and velocity in the mixed layer (and the shear that this implied at the mixed-layer/pycnocline interface). Our initial measurements from the AIDJEX Pilot Experiment demonstrated convincingly that the IOBL was not "slablike" but exhibited definite and predictable shear in the IOBL. McPhee and Smith (1976, their Figs. 8.11 and 8.12) included an example during a storm where on the second day, the Ekman layer was confined to levels well above the obvious pycnocline established by stronger forcing on the first day. Nevertheless, it remains an article of faith among many oceanographers that inertially oscillating slabs are a primary mechanism by which mixed layers remain mixed. In Section 8.2 we look at this from the perspective of a model forced with different boundary conditions.

Finally, in Section 8.3 we examine a time from the MaudNESS project near Maud Rise in the Atlantic sector of the Southern Ocean, when the cold upper layer was very close to the same density as the underlying Warm Deep Water, despite being less saline. Here we use the model to illustrate how nonlinearities in the equation of state, and possibly, differences in thermal and saline diffusion, come into play.

8.1 Diurnal Heating Near the Solstice, SHEBA

During the SHEBA drift, in June 1998, there was a period of about four days with relatively steady and moderate winds, during which ice drifted at just under 2% of the wind speed (Fig. 8.1a). This was reflected in moderate friction velocities in the upper part of the well mixed layer (Fig. 8.1b), more or less typical of average conditions for the entire project. What was notable about this period, however, was a clear, albeit small, diurnal signal in temperature of the well mixed layer, lagging solar zenith by a few hours (Fig. 8.2a). Apparently there was enough solar energy making its way through the ice cover to warm the well mixed layer during local afternoon, with some of the excess heat lost to the ice via upward heat flux at the interface when sun angle was low.

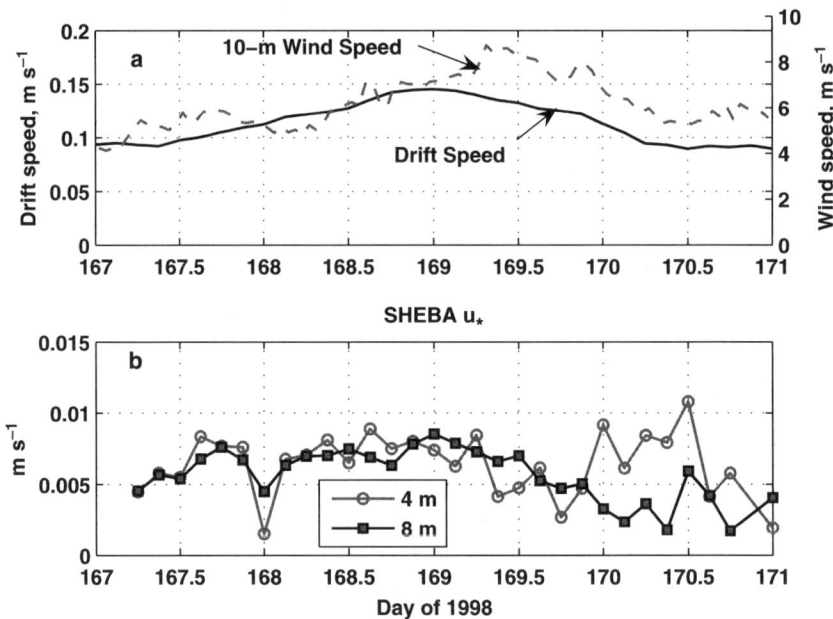

Fig. 8.1 a Wind speed at 10 m and ice drift speed after removing inertial component, 16–20 June 1998, at the SHEBA station. Ice speed ordinate range is 2% of the wind speed range. Time is shown as days of 1998, where 167.0 is 0000UT on 16 June. **b**. Friction velocity (square root of kinematic Reynolds stress) measured at two distances from the ice/water interface (see also Plate 21 in the Colour Plate Section)

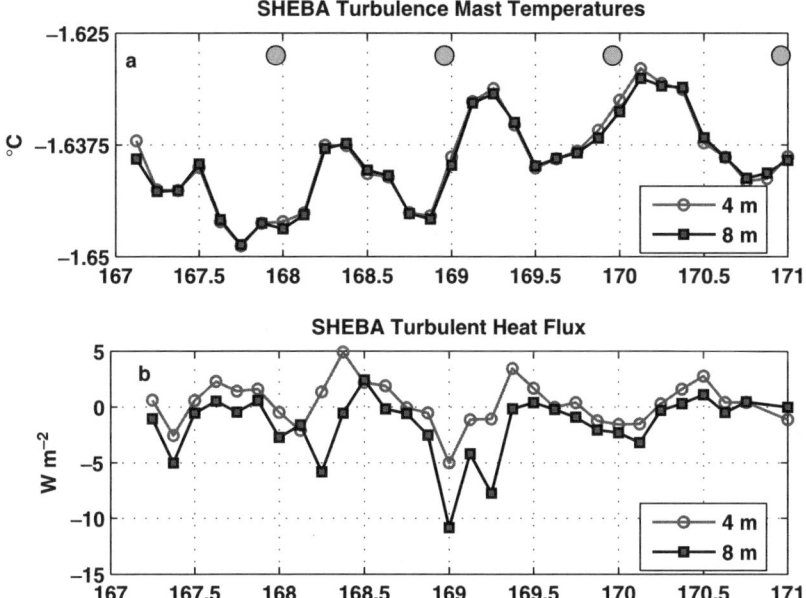

Fig. 8.2 a Three-hour average turbulence mast temperatures. Shaded circles represent local solar zenith, at approximately UT + 23 h. **b** Corresponding heat flux measurements: $\rho c_p \langle w'T' \rangle$ (see also Plate 22 in the Colour Plate Section)

The general picture of diurnal heating and nocturnal cooling was supported by turbulent heat flux measurements at the TIC levels (Fig. 8.2b). Although not so clean as the temperature records, heat flux also showed a diurnally varying signal with maximum downward (negative) flux at or shortly after local noon, and upward heat flux at night (solar nadir). There was an upward overall trend in temperature over the four days, typical of SHEBA during the early summer. This was consistent with the increase in temperature elevation above freezing $(\Delta T = T - T_f(S))$, shown in Fig. 8.3a, suggesting that the trend resulted from local heating rather than advection of the ice station into a different water type. Incoming shortwave radiation was strong during this period (Fig. 8.3a), reaching a maximum of about 600 W m^{-2} late on day 168, but with significant day-to-day variation.

There were times during the period when heat flux at both levels approached zero (e.g., days 168.75 and 169.625 in Fig. 8.2b). Again reasoning that these times would provide an accurate "calibration bath," we calculated the difference in mean temperature between the TIC sensors, which was about 1.8 mK. Adjusting the lower thermometer by this amount for all the samples then provided an estimate of the temperature gradient between 4.2 and 8.2 m as a function of time. Comparison of the negative temperature gradient with the average heat flux from the two TICs shows the time series to be well correlated, despite the small magnitudes of both (e.g., a maximum absolute temperature difference between the two SBE thermometers of

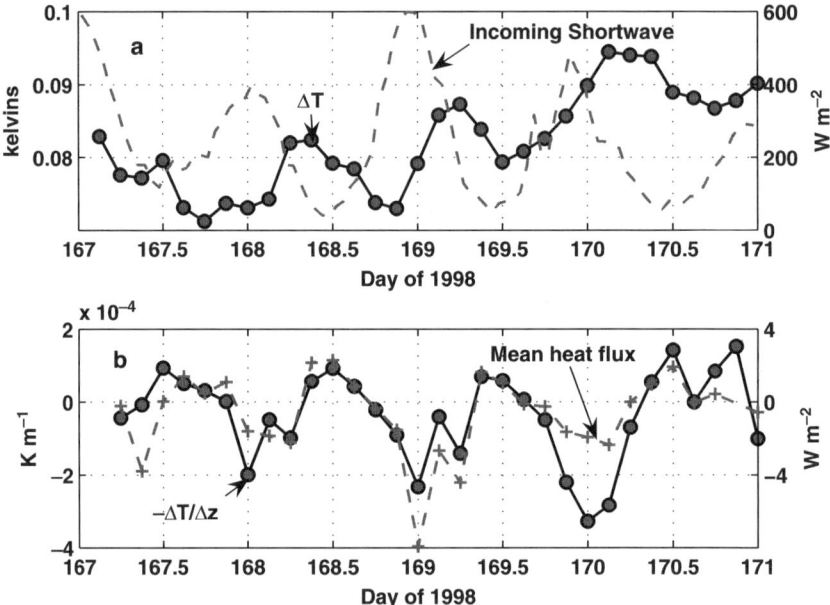

Fig. 8.3 a Incoming shortwave radiation at the upper ice surface (data from the SHEBA Project Office installation, right caption) and departure of mast temperature from freezing (average of clusters 1 and 2 at 4.2 and 8.2 m, respectively). **b** Negative temperature gradient between clusters 1 and 2, after adjusting temperatures to agree at times near zero heat flux, along with turbulent heat flux averaged for both clusters (see also Plate 23 in the Colour Plate Section)

about 1 mK).[1] To the degree that the different scale limits in Fig. 8.3b represent in the same way the variation in $-\partial T/\partial z$ and $\rho c_p \langle w'T' \rangle$, their ratio provides a rough estimate of the mean eddy diffusivity throughout the period, namely about $2.5 \times 10^{-3} \, \text{m}^2 \, \text{s}^{-1}$.

The period from day 167 to 171 (16–20 June 1998) was modeled as follows. Initially the temperature and salinity of the upper ocean were set to values measured by the SHEBA profiling CTD using a 3-h average centered at time 167.0. The upper boundary condition for momentum flux was specified (Section 7.2.2) from the time series of ice velocity after removing inertial motion. Surface roughness was assigned a value of 0.048 m from the "scaled up" analysis described later in Section 9.3.3. This is significantly rougher than the estimate for the immediate SHEBA Site 2 roughness (McPhee 2002), but was thought to be more appropriate for modeling thermal changes over the entire upper ocean. For the interface submodel, heat conduction in the ice was estimated from the temperature gradient in the lower 50 cm of ice at mass balance station "Pittsburgh," taken to be fairly representative of the entire floe. Over the four-day period, the average upward heat conduction in the ice based on this gradient was $\sim 1.9 \, \text{W m}^{-2}$.

---

[1] For very small gradients, the adiabatic lapse rate in the well mixed layer is important; however, by calibrating the thermometers to a time of near zero heat flux, the difference between temperature and potential temperature has already been taken into account.

8.1 Diurnal Heating Near the Solstice, SHEBA

Solar radiation was introduced into the model water column by taking a fixed fraction (f_{SW}) of the incoming surface solar radiation (I_0) measured at the SHEBA Project Office site, and distributing it with an exponential attenuation in the upper ocean with an e-folding depth $\ell_{SW} = 4$ m, to generate a source term

$$Q^H = \frac{f_{SW} I_0}{\ell_{SW}} e^{\frac{z}{\ell_{SW}}}$$

Parameters important in the interface submodel were assigned values: $\alpha_h = 9.3 \times 10^{-3}$, $\alpha_s = \alpha_h/35$, and $S_{ice} = 6$ psu.

The main point of the modeling exercise was to gauge how much of the incoming radiation measured at the surface had penetrated the ice cover, in order to realistically account for both the diurnal variation and the secular trend in upper ocean temperature. The model results, run with three different values of f_{SW} as shown in Fig. 8.4, indicate that about 8–10% of the short-wave radiation made its way through the multiyear ice floe. With these fractions, the model results indicate about the right amount of total heating over the four-day period (Fig. 8.4a), and show the same pattern of diurnal variation, although the model appears to somewhat underestimate the nocturnal cooling. Modeled friction velocity at 8 m is fairly accurate; mean values differ by less than 0.2 mm s^{-1}. At 4 m (not shown) average modeled u_* exceeds measured by about 0.5 mm s^{-1}, perhaps not surprising because z_0 in the model is substantially larger than estimated for the smooth ice surrounding Site 2 (McPhee 2002).

Downward heat flux near maximum sun angle is modeled reasonably well at 8 m (Fig. 8.4c); however, the nocturnal upward (positive) heat flux is larger in the model than observed. At night the main heat sink in the system is basal melting, which in the model is consistent throughout the simulation period, so this might suggest that modeled basal heat flux is too large (i.e., α_h is too large). This interpretation, however, is at odds with the relatively rapid observed nocturnal cooling (compared with the models) in Fig. 8.4a. A more likely explanation is that ice in the vicinity of the turbulence mast is smooth enough compared with the overall roughness of the floe (and model with $z_0 = 4.8$ cm) that the heat extraction from the water column is smaller than the average for the whole floe. Temperature of the well mixed layer would represent an integrated effect.

For the model run with $f_{SW} = 0.09$, the mean basal heat flux was about 17 W m^{-2}, which, when reduced by the small conductive flux in the ice, implied ice melt of around 2 cm over the four days. This produced an average buoyancy flux at the interface, $\langle w'b' \rangle_0 \approx 10^{-8}$ W kg^{-1}, which had little effect on turbulence near the surface, but produced a mean value for μ_* of about 2.7 (Section 4.2.3). This is enough to reduce the dynamic boundary layer extent slightly (see Fig. 4.8).

When we first did these simulations, we were surprised that as much as 10% of the incoming solar radiation was making its way into the water. At the time, estimates from aerial photography put the fraction of ice area covered by open leads in the SHEBA region at about 2.5% (Perovich et al. 2002, their Fig. 8.6), providing effectively a lower limit on f_{SW}. By mid June melt ponds had formed and were estimated to cover 15–20% of the surface (Perovich et al., op. cit.). Melt ponds

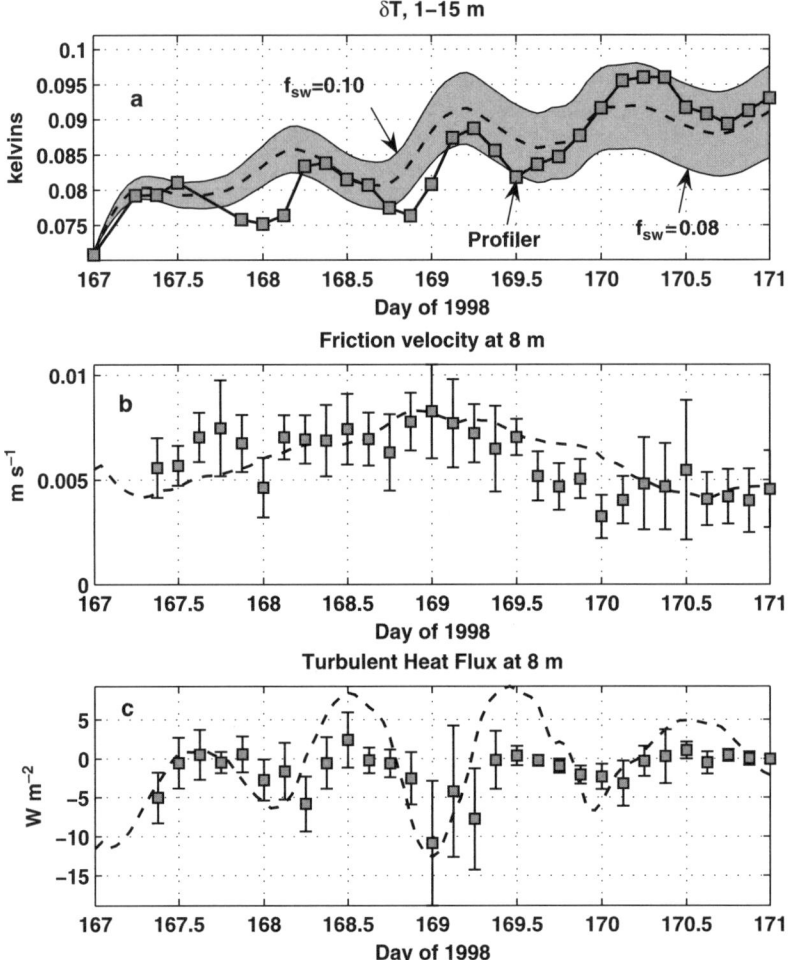

Fig. 8.4 a Comparison of model well mixed layer temperature elevation above freezing in the upper 15 m of the water column with results from the SHEBA profiler. Dashed curve is for $f_{SW} = 0.09$, envelope limits are shown. **b** Modeled (dashed curve, $f_{SW} = 0.09$) and observed friction velocity at 8 m (error bars indicated ± one standard deviation of the 15-min realizations in each 3-h average). **c** Modeled and observed turbulent heat flux at 8 m

typically reduce the surface albedo significantly and provide a potential conduit for short wave radiation entering the ocean. Still, discussions with G. Maykut (2000, personal communication) and others implied that, given the highly compact sea ice cover in the SHEBA region, it was unlikely that much more than 4% of the incoming radiation would be able to transit the ice pack, given commonly accepted values for shortwave broadband ice extinction coefficients (e.g., Grenfell and Maykut 1977). This was half or less than the amount of heating needed to produce the observed diurnal variation. The possibility of the ocean measurements being contaminated by the presence nearby of open water remained, but analysis of aerial photography indicated that any open water "upstream" of the ocean measurements was far enough away to have had little impact. The apparent discrepancy between the modeling example here and earlier estimates of light transmittance in ice appear to be resolved by recent work of Light et al. (2008). They report extinction coefficients for bare and ponded ice at SHEBA that are substantially smaller than previous estimates, and that 3–10 times as much solar radiation penetrates the ice cover than is predicted by current global circulation models.

8.2 Inertial Oscillations in Late Summer, SHEBA

By mid September at the SHEBA station in 1998, the ice cover was relatively compact, the well mixed layer had cooled to within few centikelvins of freezing but remained relatively shallow, as evidenced by strong inertial oscillations during much of the month (Fig. 8.5). Usually, the presence of strong inertial oscillation signals that the internal ice stress gradient is small enough that ice is in a "free-drift" state, i.e., wind driven. We chose a period during 14–22 September 1998, as a sort of "modeling laboratory" to look as different aspects of the upper ocean response (both modeled and observed) during a period of significant inertial oscillation. At 0900 on 14 September (257.375), the wind was still and ice drift speed near zero. Over the next five days, ice drift speed rose rather steadily to about 0.3 m s^{-1}, following wind closely at slightly over 2% of the wind magnitude (Fig. 8.5c). Late on day 262, wind dropped quickly, as did mean ice drift; however, a fairly strong train of inertial oscillations continued for several days.

8.2.1 Wind Forced Model

The first LTC model exercise (SEP 14A) was initialized with the SHEBA profiler 3-h average *T/S* data in the upper 61 m of the water column, and driven by wind stress for the period 257.375–265, obtained by applying a drag coefficient, $c_{10} = 0.002$, to the observed 10-m wind at the project office tower, and using the dynamic boundary condition for stress (Section 7.2.3). During this period shortly after the start of freeze-up, the temperature gradient in the lower part of the ice was slightly positive, indicating that even in relatively thin ice at the end of the melt season, the downward "freezing wave" had not yet reached the ice base. This was incorporated into the model as a *downward* interface heat flux averaging about -0.6 W m^{-2}. Since there was not a lot of open water, for simplicity, we assumed that the long wave radiative loss from open water would roughly cancel incoming short wave gain. Undersurface roughness was set at 0.048 m as above.

Modeled versus observed surface velocity (Fig. 8.6) demonstrates reasonable simulation of both the mean and inertial components of ice velocity, including

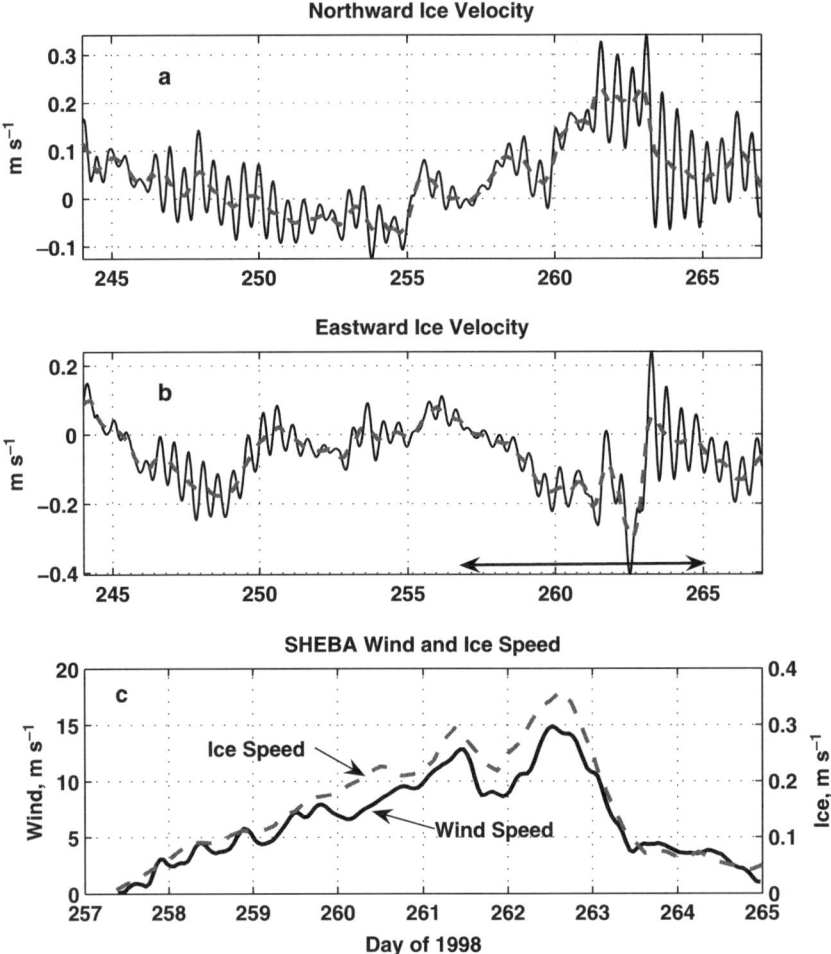

Fig. 8.5 a Northward ice drift velocity during September of the SHEBA project from satellite global positioning data. **b** Eastward component. The arrow indicates time shown in: **c** Wind speed and drift speed during the time from 257.375 (0900 UT on 14 September 1998) to 265.0. The scale for ice speed is 2% of the wind speed scale (see also Plate 24 in the Colour Plate Section)

the energetic oscillations associated with the rapid change in drift direction from NW to N (see Fig. 2.3). Reynolds stress measurements are compared with model estimates in Fig. 8.7. In general, they follow reasonably well, although on day 259 friction velocity 6 m from the boundary is inconsistent with both the 2-m measurements and the model. This may be due to "upstream" disturbance in the underice morphology from the particular drift direction on that day. Modeled and measured turbulent heat flux (Fig. 8.8) are also reasonably matched, although the model underestimates heat flux at 2 m on day 262. The overall assessment is that the relatively simple, first-order closure model driven only by wind and initial T/S conditions is successful at simulating the main features of both surface velocity (including inertial oscillation) and upper ocean turbulent fluxes.

8.2 Inertial Oscillations in Late Summer, SHEBA

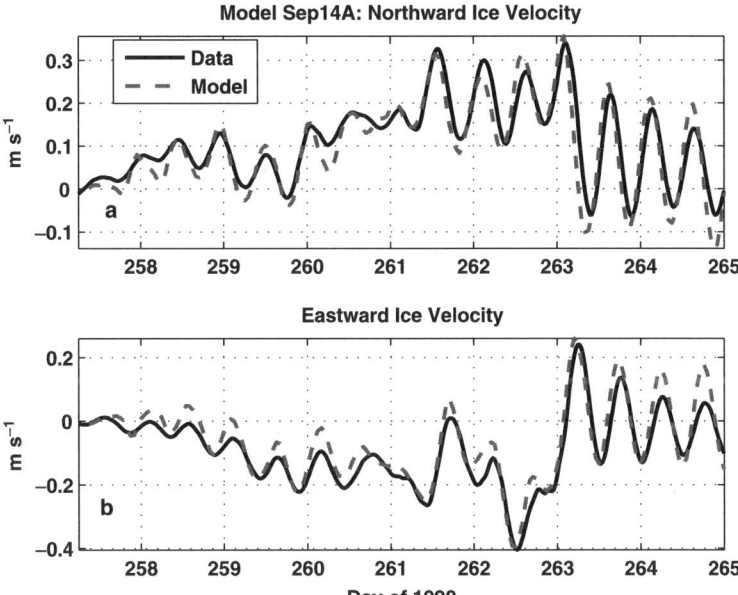

Fig. 8.6 Observed (solid) and modeled (gray dashed) ice velocity northward **a** and eastward **b** in the period from 0900 UT 14 September 1998 to 0000 UT 22 September, for the wind driven model run Sep 14A

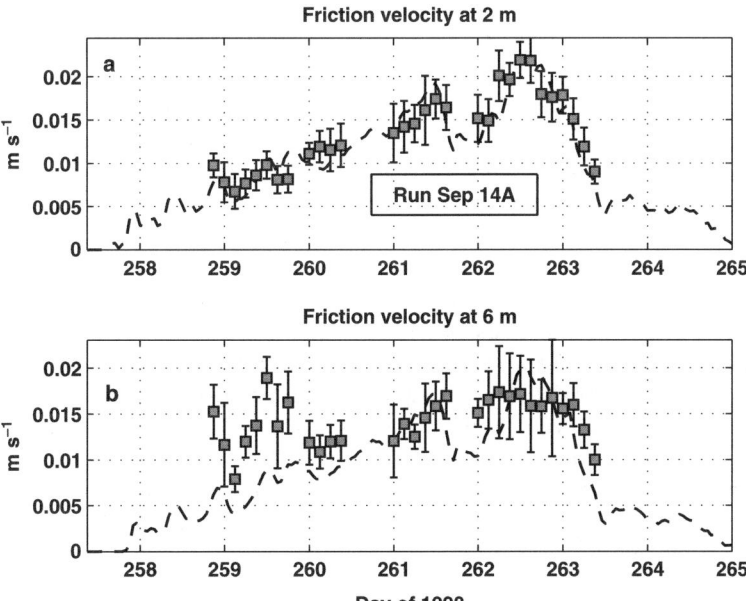

Fig. 8.7 Modeled (dashed) and observed (square symbols, $u_* = |\langle u'w' \rangle + i \langle u'w' \rangle|^{1/2}$) for model run Sep 14A, at 2 m **a** and 6 m **b** from the boundary. Error bars represent ± one standard deviation of the 15-min samples in each 3-h average

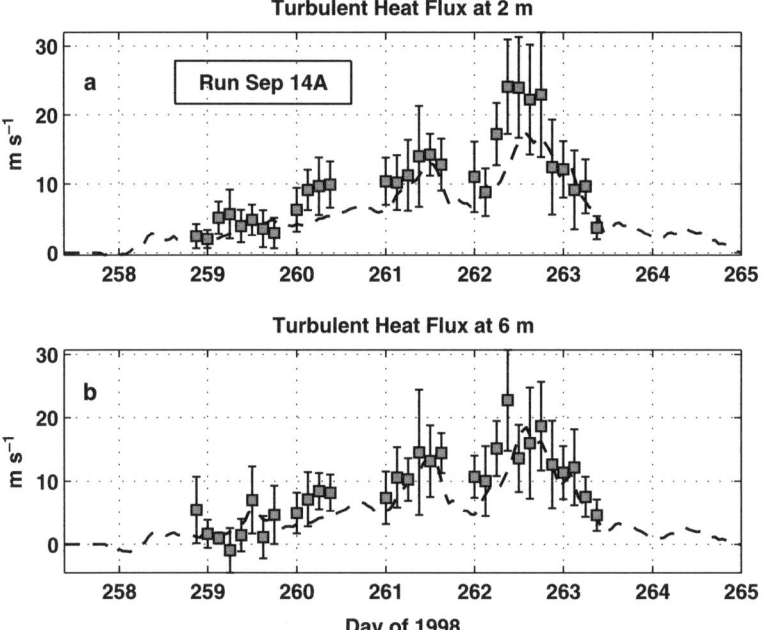

Fig. 8.8 As in Fig. 8.7, except turbulent heat flux, $\rho c_p \langle w'T' \rangle$

It is not nearly as successful, however, at reproducing the mean evolution of the upper ocean structure. In the model, for example, the well mixed layer salinity (Fig. 8.9a) increases substantially in response to strong surface stress from 261–263 as more saline water is mixed upward from the pycnocline. Observed salinity decreases. In the model, δT also increases by upward mixing of warmer water, counter to the downward trend in the observations. In the absence of other information, it would appear then that the model has underestimated the amount of melting, since in a one-dimensional view, melting is the only source of fresh water, and more rapid melting would lower δT. But contour plots of salinity from the profiler and the model point up some other major differences. In the model, the impact of mixing is minor below about 40 m and salinity in the lower part of the model domain remains the same, while in the data there is an overall freshening trend (downward sloping isohalines) with an upwelling-like event during the time of maximum stress and a rapid deepening of the well mixed layer on day 262.

In the absence of mixing from below, the only source of salt in the model domain is freezing (positive) or melting of ice. The integrated change in salt over the 60 m domain of run Sep 14A is

$$\Delta_S = \int_{-60}^{0} S_{end} dz - \int_{-60}^{0} S_{start} dz = -0.50 \, \text{kg m}^{-2}$$

8.2 Inertial Oscillations in Late Summer, SHEBA

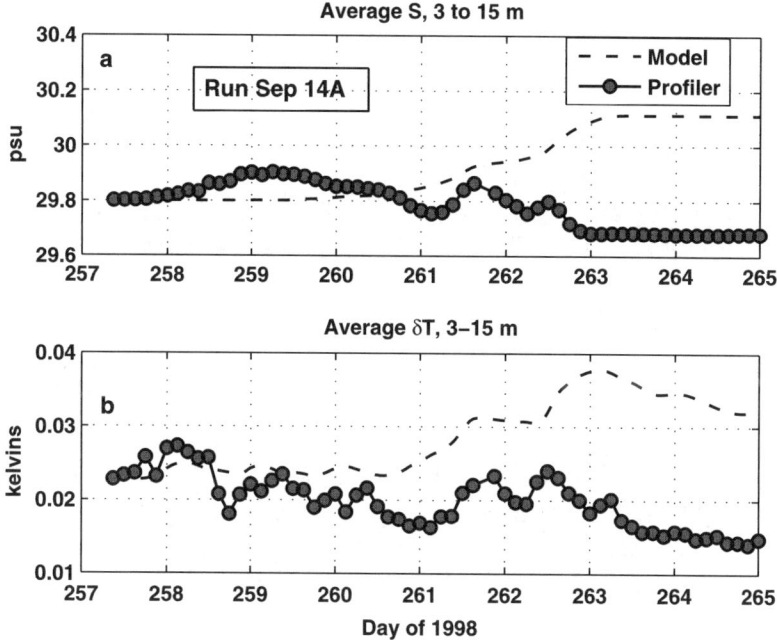

Fig. 8.9 Well mixed layer properties: **a** salinity and **b** temperature elevation above freezing. Circle symbols are 3-h averages of the SHEBA profiler data; dashed line is from model run Sep 14A

equivalent to about 2 cm of ice melt. A similar calculation for Δ_S in the upper 60 m from the starting and ending SHEBA salinity profiles shows a change in salt content of about $-36\,\mathrm{kg\,m^{-2}}$, requiring enough melting (1.7 m) to completely eliminate the ice pack! Another model simulation (run Sep14B) was made in which a small constant advective source term was specified across the entire model domain at each time step:

$$\dot{S} = \frac{\Delta_{S(obs)}}{H(t_{end} - t_{start})}$$

where $\Delta_{S(obs)}$ is the observed total change in salt content in the upper 60 m over the ∼8 days, making the modeled Δ_S match the observed value. The model results for stress are not much different from run Sep 14A; modeled heat flux is somewhat less. Results for δT in the well mixed layer (Fig. 8.11) show about the same overall decrease as in the data. The simulation is not meant to be very realistic, given the temporal changes evident in Fig. 8.10a, but rather to show that the observed decrease in δT probably resulted from advection instead than local vertical processes. In effect, the advective decrease in salinity as the station drifted north more than offset the upward vertical mixing of both salt and heat from the pycnocline.

An important point to be made from this exercise is that in general it is quite difficult to evaluate the performance of upper ocean models by testing their ability to simulate short-term changes in mean properties of the upper ocean. Often

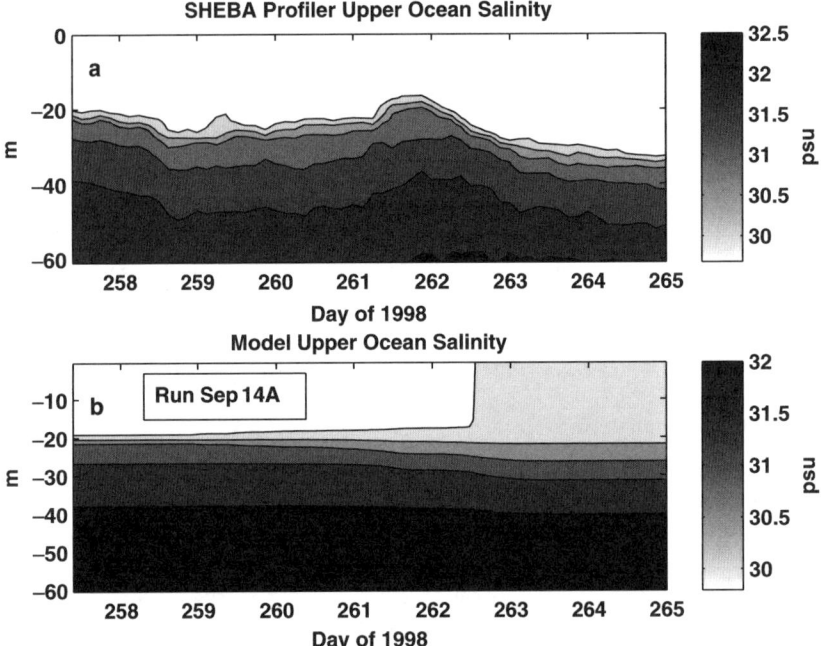

Fig. 8.10 Contour plots of salinity in the upper ocean from SHEBA profiler **a** and wind-driven model Sep 14A **b** (see also Plate 25 in the Colour Plate Section)

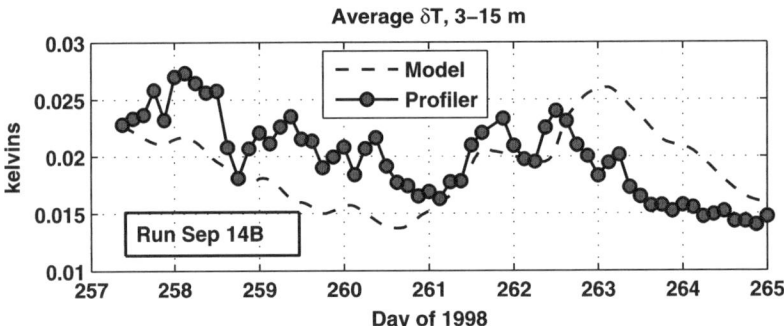

Fig. 8.11 Comparison of modeled and observed dT for model Sep 14B with a small negative salinity source term as an approximation to advective flux

relatively small horizontal gradients encountered as the ice drifts will swamp the signal from local vertical exchanges. On the other hand, vertical fluxes are often relatively immune to horizontal property gradients except in front-like conditions, hence in general provide a superior assessment of the particular mixing scheme used in the model.

8.2.2 Models Forced by Surface Velocity

A third model simulation of the same period (run Sep 14C) is identical to run Sep 14A except that the dynamic stress interface boundary condition was replaced by the surface velocity boundary condition as in Section 7.2.2. The same value (0.048 m) was used for z_0. Apart from minor details the two simulations are similar for exchanges at the interface (Fig. 8.12) as well as the rest of the IOBL. The mean modeled values for friction velocity over the simulation period differ by about 10% (if c_{10} is increased to 0.0025 they match). During winter there are often periods when the ice clearly responds to wind forcing, but is also influenced by internal stress gradients. In this case, provided the geostrophic (sea-surface tilt) velocity is small compared to ice velocity, forcing the IOBL with ice velocity clearly is a better strategy than forcing by wind (unless ice stress is known).

Especially during winter, a common consequence of internal ice forcing is that inertial oscillation is severely damped even if the ice appears to be moving freely in response to the wind (McPhee 1981). An obvious question then arises: would the modeled response of the upper ocean be much different if inertial oscillation is absent? We performed a fourth simulation (run Sep 14D) of the September 14–22 period which was identical to run Sep 14C except that surface velocity was specified

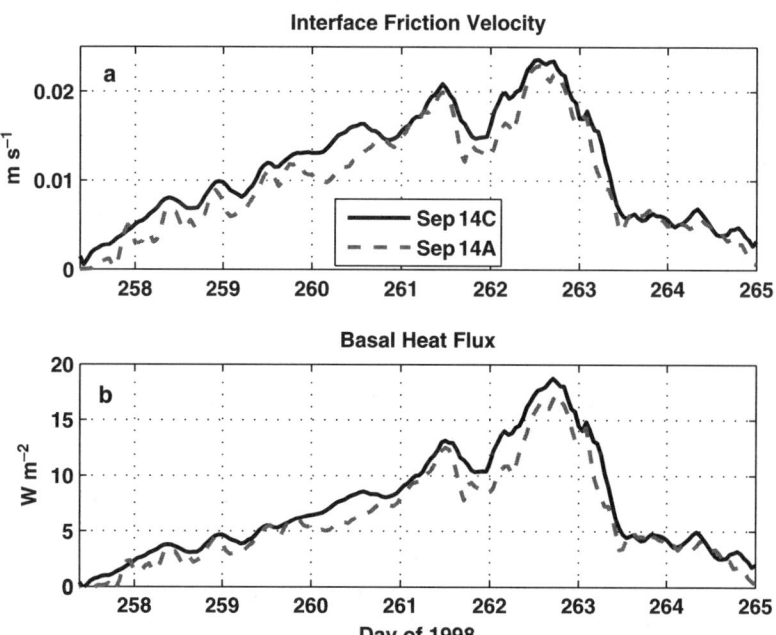

Fig. 8.12 Comparison of interface friction velocity **a** and basal heat flux **b** for model run Sep 14C (forced by surface velocity) and Sep 14A (forced by 10-m wind)

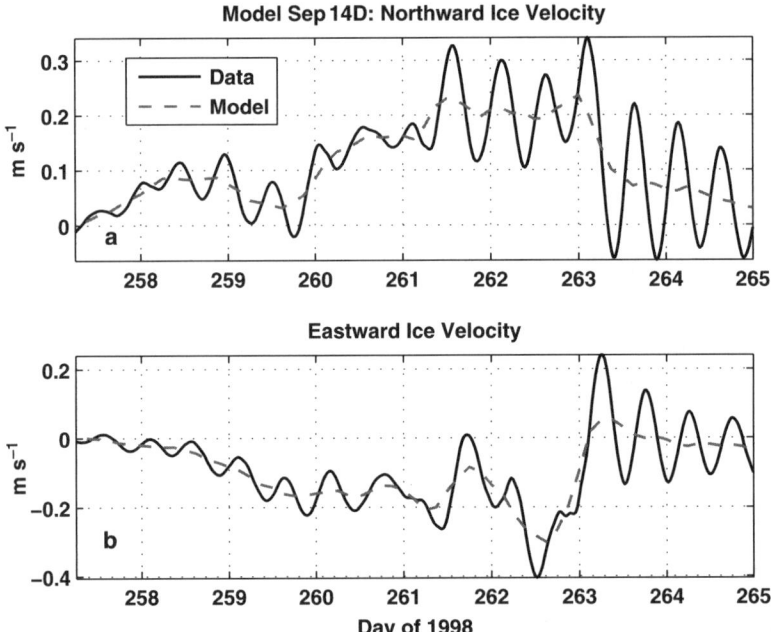

Fig. 8.13 Velocity (dashed) used to force the model run Sep 14D, compared with the actual ice velocity. **a** Northward component; **b** southward component

as the ice velocity after removing inertial components (i.e., forced by the V_0 term in [2.22]). Surface velocity used to force the model is shown as the dashed curves in Fig. 8.13. Differences in results from the models forced by the complete surface velocity (run Sep 14C) and by surface velocity with inertial components removed are summarized at two levels in Figs. 8.14 and 8.15. The ML/Pycnocline level is defined as the deepest z (flux) grid point in the well mixed layer, i.e., where

$$N^2 = -\frac{g}{\rho}\frac{\partial \rho}{\partial z} \leq 1.5 \times 10^{-5}\,\text{s}^{-2}$$

It varies with time, but because of the strong initial stratification, remains relatively shallow, averaging about 15.4 m for each run.

That there is little difference in u_{*0} between the models is not surprising (Fig. 8.14a) — we seldom see much inertial component in velocity measured near the ice (in a reference frame drifting with the ice), because the ice and upper ocean oscillate in phase, and mean shear is not much affected. Near the base of the well mixed layer, it is not obvious that inertial shear would be so unimportant, but according to the model comparisons (Fig. 8.14b), the impact remains relatively small. Similar results hold for the turbulent heat flux. There is some reduction in mean heat flux when inertial oscillation is removed: about 6% near the base of the well mixed layer, and 4% at the interface. Nevertheless, it appears that even with

8.2 Inertial Oscillations in Late Summer, SHEBA

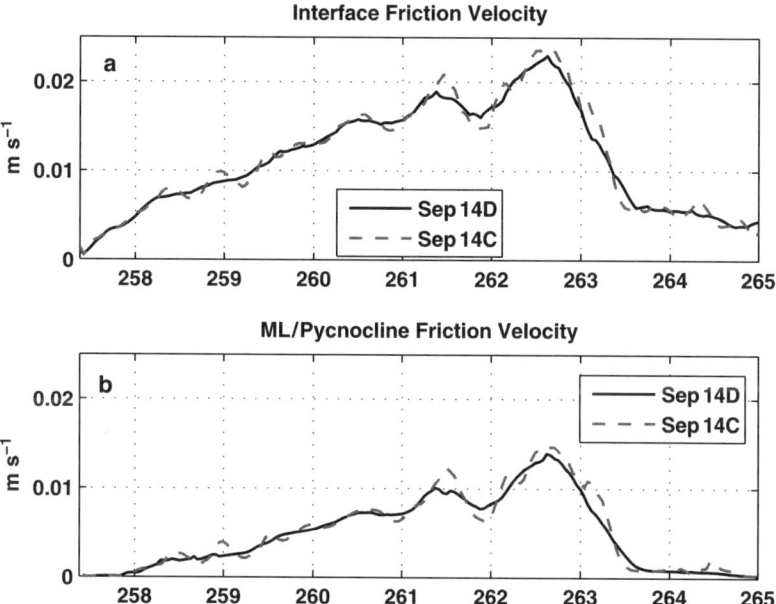

Fig. 8.14 **a** Comparison of u_{*0} as modeled with total-velocity forcing (Sep 14C, dashed gray curve) and without inertial motion (Sep 14D, solid black). **b** Same as **a** except at the lowest grid point in the well mixed layer as explained in the text

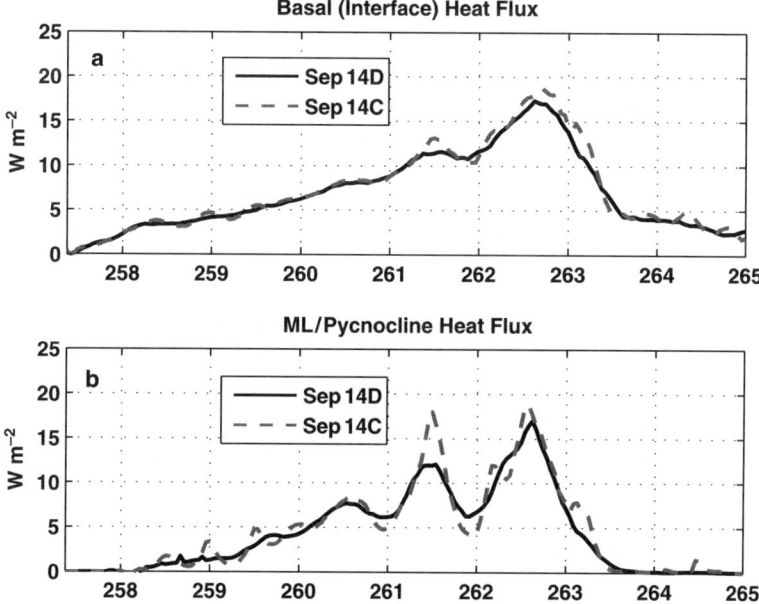

Fig. 8.15 Same as Fig. 8.14, except turbulent heat flux

energetic inertial motion and a relatively shallow pycnocline, inertial shear may play a relatively minor role in upward mixing. For the SHEBA circumstances, the pycnocline is very strong. For weaker stratification near the base of the mixed layer (as in the Southern Ocean), inertial shear may be significantly more important in the mixing process.

8.2.3 Short-Term Velocity Prediction

Development of the complex demodulation technique for analyzing sea-ice drift described in Section 2.5 (see also McPhee 1988) was spurred by a practical task of predicting where ice floes tracked by radar in the vicinity of exploratory oil drilling platforms in the Beaufort Sea might drift over the next few days. Inertial oscillations were a prominent feature in the radar tracked trajectories of nearby floes, with rapidly changing directions and speeds, so that depending on where the tracking picked up in the inertial loop, the floe might be headed directly toward the platform at high speed, then a short time later going in a quite different direction. Extrapolating future drift from a short history of observed drift thus required consideration of the inertial motion.

The problem of starting a model at a particular time for short term predictions is related to the fact that a simple harmonic oscillator (e.g., equation (2.22)), when forced from rest impulsively will oscillate continuously about a steady state that it never reaches. In reality, of course, the ice/IOBL system is not frictionless, yet it is clear from records like Fig. 8.5 that oscillations can persist for several days. In the SHEBA examples of Sections 8.2.1 and 8.2.2, the model was started from a time when wind and ice drift velocity were nearly zero, so impulsive initial forcing was not much of an issue. But suppose we wished to predict ice motion starting early on day 263, when the wind is high but forecast to diminish, and inertial motion is large. To highlight the problem, we drive the ice/IOBL system with the observed wind, initialized with T/S structure as observed at time 263.125, and started from rest (all velocities zero). Results for drift velocity (Fig. 8.16) show that although inertial oscillations are generated in the model, they are substantially out of phase with the observations and would be of little use in actually estimating where the floe would be in a short time.

To address this problem the IOBL model was initialized by (i) solving a steady version of the model (described in detail in Chapter 9), and then (ii) adding to the steady solution for velocities in the mixed layer, the inertial component of velocity from the complex demodulation record synthesized from the GPS positions (this assumes that the ice and IOBL are oscillating in phase). In this case (Fig. 8.17), the first few inertial cycles are much closer to the observations.

8.2 Inertial Oscillations in Late Summer, SHEBA

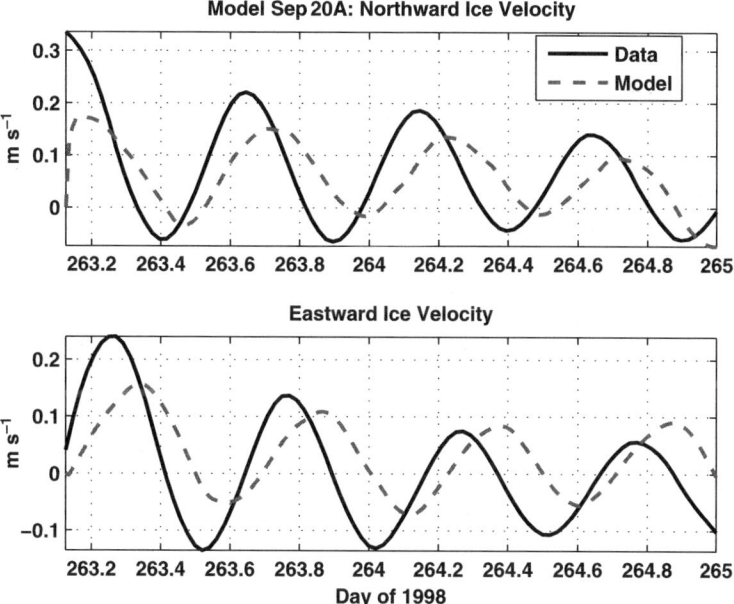

Fig. 8.16 Simulated surface velocity compared with observed for a model forced by observed wind and started from rest at a time of high wind stress and large inertial motion. The model quickly adjusts to the stress forcing, but generates its own inertial oscillations out of phase with the observations

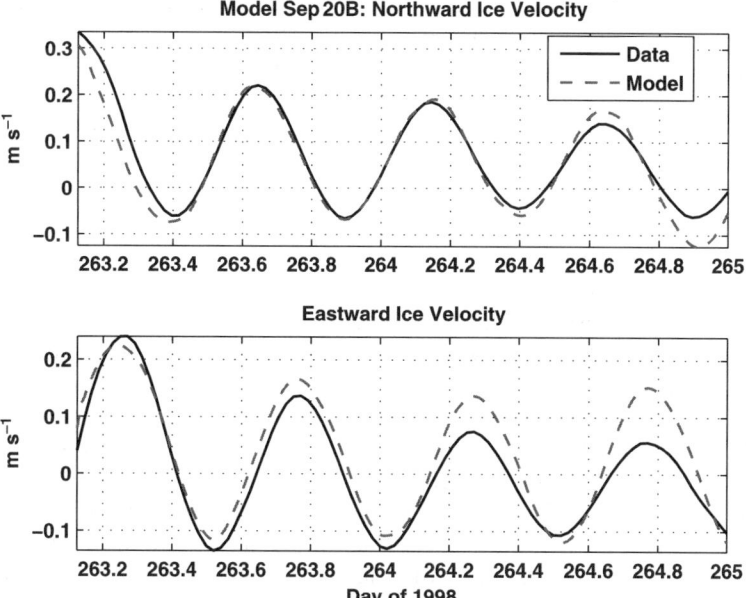

Fig. 8.17 Results from a model identical to that of Fig. 8.15 except that the initial velocity structure is specified by solving a "steady" version of the IOBL model given the wind stress at time 263.125, and adding the inertial component of velocity from satellite navigation analysis to the well mixed layer velocities

8.3 Marginal Static Stability, MaudNESS

The last example of applying the one-dimensional LTC model arises from another practical requirement encountered in planning for the MaudNESS experiment near Maud Rise in the Weddell sector of the Southern Ocean. The basic plan for MaudNESS was to perform a fast, relatively shallow CTD survey across the seamount, concentrating on the margins, and use a combination of special weather forecasts and ice concentration analyses to estimate the most likely regions for thermobaric instability and deep-reaching convection based on CTD stations made at different times and places. To this end a forecast model was needed that was simple enough to update several candidate profiles with every new (daily) weather forecast, with the goal of enhancing operational planning by identifying areas where mixing would occur in the least stable density environment. Overall, our objective was to measure turbulent exchange in a low stability environment to understand what conditions might remove the thin ice cover completely over a substantial area.

A comparison of upper ocean conditions during late summer at SHEBA in the western Arctic (79.9°N, 161.6°W) with late winter at MaudNESS in the Weddell (65.5°S, 001.1°E) illustrates a striking contrast in static stability of the water column (Fig. 8.18). In the former (used to initialize the model run described in Section 8.2 above), potential density (Fig. 8.18c) increases by nearly 3.5kg m^{-3} in the upper 150 m, while for the latter (MaudNESS Station 91), the increase is two orders of magnitude less, about 0.03kg m^{-3}, and is barely perceptible when drawn at the same scale. Thus by comparison the Weddell profile is near to being statically neutral; nevertheless, there is a steep thermocline (Fig. 8.17a) starting at around 100 m with far more heat content close to the surface than in the Arctic. A much less obvious halocline (again when drawn at the scale appropriate for the Arctic) contributes to a slightly stable pycnocline (in potential density, less stable for *in situ* density) that separates the upper cold layer from the underlying Weddell Deep Water. Because of the low stability, it is relatively easy to mix heat upward from the large WDW reservoir, whenever there is vigorous stirring at the surface (which is common in the Weddell in winter). But this is the source of the *thermal barrier* (Martinson 1990) that melts ice when basal heat flux exceeds conduction through the ice cover, forming a shallower halocline that severely inhibits deeper convection. As the profiles indicate, the system is very delicately balanced. Note that below about 120 m in the MaudNESS profile, temperature and salinity are very uniform, indicating some active mixing activity.

In the operational mode, the model strategy was to use the weather forecast and ice concentration data provided daily via satellite communications from the Arctic Mesoscale Prediction System (AMPS) MM-5 regional model (Powers et al. 2003) to project ahead five days from the current time, keeping backward track at a particular location from the time at which an initial upper ocean structure was measured. This was done by accepting the "nowcasts" that initialized each daily weather forecast as valid analyses. Comparison of the nowcasts with ship observations were generally favorable, although the MM5 temperatures often appeared to be biased high by a few degrees. In this way, we could construct for a given location in the operation

8.3 Marginal Static Stability, MaudNESS

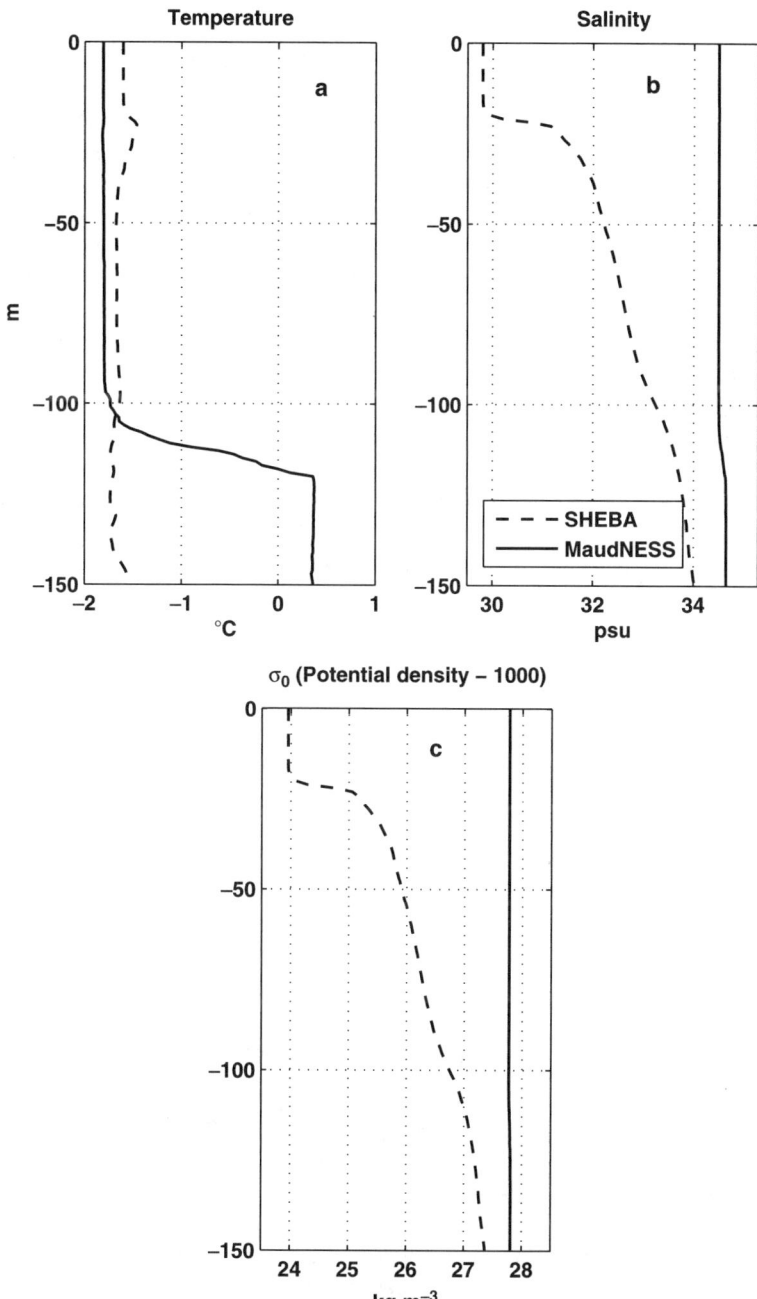

Fig. 8.18 Comparison temperature **a**, salinity **b**, and σ_0 **c** profiles in the upper 150 m from SHEBA in the western Arctic on 14 September 1998 (dashed) and from MaudNESS in the Weddell Sector, Antarctica, 19 August 2005 (solid)

area a continuous record of wind, temperature, ice concentration, and other pertinent fields. In this section, we use such a time series to model the subsequent upper ocean conditions starting from observations about noon UT on 19 August 2005 at Maud-NESS Phase 1 station 91, and extending 20 days. Station 91 was chosen because it exhibited the lowest "thermobaric barrier" index (McPhee 2000) of the MaudNESS phase 1 survey. The 1-d ocean forecast model accounts for open water as stipulated in the supplied ice concentration data (averaged in a 90 km square centered on the station site) by assuming that when ice is absent, the surface loses $200\,\mathrm{W\,m^{-2}}$, and that, provided this exceeds heat flux from ocean, it forms ice that migrates away from the region (but leaves salt). The treatment is crude, but proved not very important since according to the imagery analysis, ice concentrations remained high during the modeling period. Heat loss through the ice was estimated by assuming a one-layer ice model with heat flux proportional to the air-ocean temperature difference divided by the ice thickness. Thermal conductivity in the ice was assumed to be $2\,\mathrm{W\,m^{-1}\,K^{-1}}$. Surface momentum flux was estimated from MM5 surface wind with a 10-m drag coefficient 0.0015. The time series synthesized from the MM5 nowcasts at the Station 91 site is summarized in Fig. 8.19. Throughout the period, winds were moderate (by Southern Ocean standards) with an average temperature around $-14\,^{\circ}\mathrm{C}$.

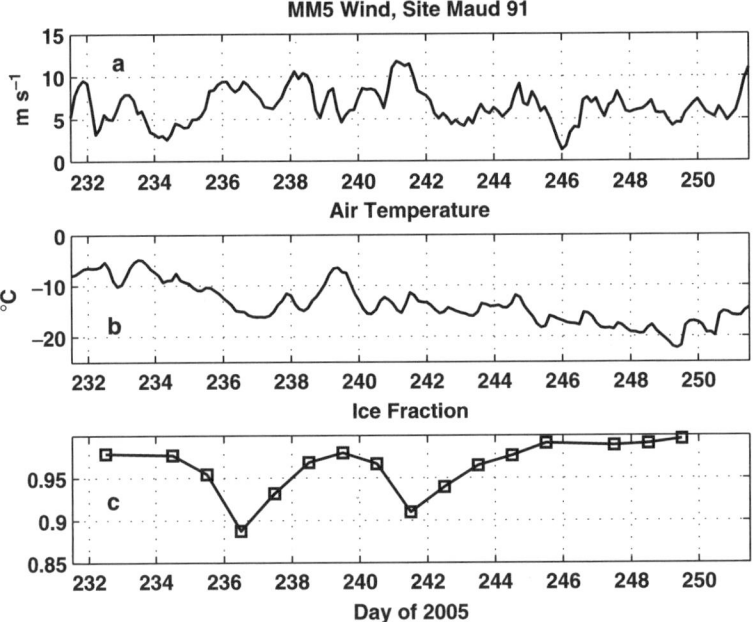

Fig. 8.19 Environmental parameters extracted from the MM5 model output: **a** 10-m wind speed; **b** 2-m air temperature; and **c** ice concentration

8.3 Marginal Static Stability, MaudNESS

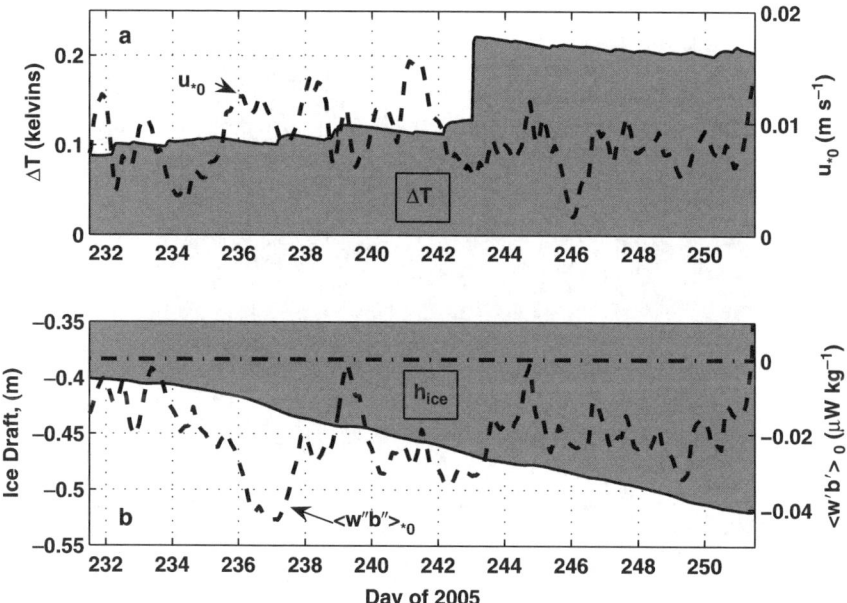

Fig. 8.20 **a** PD Model interface friction velocity (dashed) and departure of mixed layer temperature from freezing (shaded); **b** ice draft (shaded) and interface buoyancy flux (dashed) for an IOBL model where all density gradients are based on potential density. (see also Plate 26 in the Colour Plate Section)

In the modeling examples described below, the intent is not to faithfully predict the evolution of upper ocean structure at station 91 for the entire 20-day period (which is unknown), but rather to explore ramifications of mixing in a low static stability environment. To that end, the model was run first with all of the buoyancy flux dependent parameters calculated from gradients in potential density, and with no mixing (beyond molecular) below the IOBL extent, which was determined by dynamic conditions at the surface along with modeled changes in the temperature and salinity profiles. This model is designated PD (indicating that gradients depend on potential density) and its results are summarized in Fig. 8.20. Parameters that govern basal heat flux (friction velocity and elevation of mixed layer temperature above freezing) show that in the second part of the period, higher ΔT is somewhat compensated by lower u_{*0} and lower air temperature, so that until the very end of the period, ice continues to grow as its upward heat conduction (plus small loss in open water) exceeds heat flux from the ocean. Total ice growth of about 12 cm supplies a small but nearly continuous negative buoyancy flux (dashed curve in Fig. 8.20b, mean value -2×10^{-8} W kg^{-3}). In the PD model, this contributes to turbulence in the IOBL and a slow deepening of the thermocline (Fig. 8.21). For the most part, the dynamic IOBL depth (white dashed curve in Fig. 8.21, defined here as the depth below which $u_* < 0.5 \times 10^{-3}$ m s^{-1}) follows the thermocline closely. As upper layer salinity increases from the downward salt flux, the pycnocline weakens, until on about day 243, it erodes enough of the thermocline to cause a noticeable temperature spike

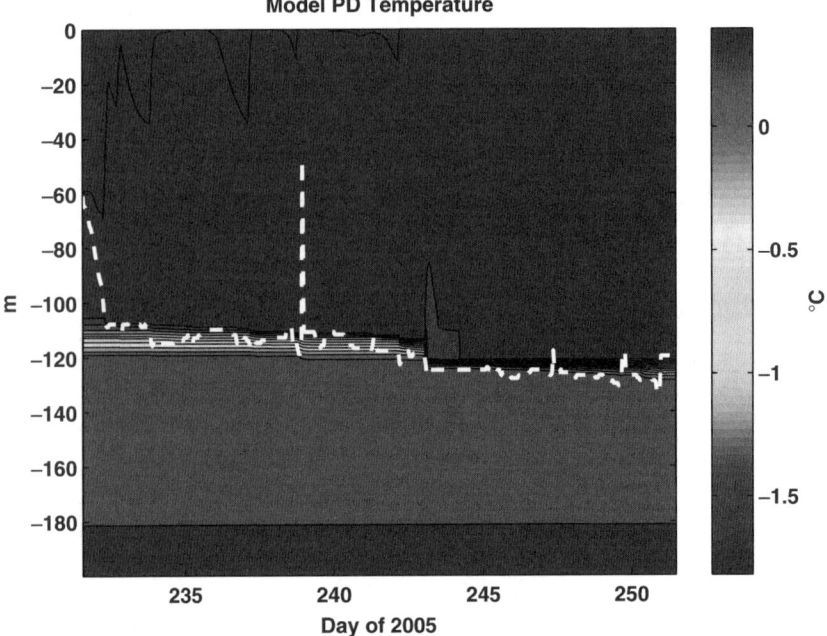

Fig. 8.21 Contours of PD model temperature for the upper 200 m of the water column. The white dashed curve is the dynamic boundary layer depth, below which friction velocity is less than 0.5 mm s^{-1}

near the base of the upper layer, and a rapid increase in mixed layer temperature by about 0.1 K (Fig. 8.20a). Note that because of the large temperature contrast, minor deepening of the well mixed layer has large impact on its temperature, thus strongly reinforcing the "thermal barrier" effect.

During the 20-day PD model run, the density contrast between the upper and lower layers continues to decrease. In Fig. 8.22, the density jump across the thermocline is shown from two perspectives, one in which it is simply the difference in potential density (dashed), and the second when it is calculated at pressure corresponding to the depth of the thermocline (solid). The shaded areas are thermobarically unstable, i.e., water just above the thermocline, if displaced slightly downward, would be heavier than its surroundings. Any subsequent mixture of upper- and lower-layer water would be denser than either type by itself by virtue of the curvature of isopycnals in T/S space (Fig. 2.11). To address this in the context of the one-dimensional LTC model, we formulated a simple algorithm as follows.

In the first-order-closure model, buoyancy flux is calculated as the eddy diffusivity times the gradient in buoyancy frequency squared, $N^2 = (-g/\rho)\rho_z$, where in the model grid scheme (Fig. 7.1)

$$(\rho_z)_i \doteq \frac{\rho_i - \rho_{i-1}}{\Delta z z_{i-1}} \qquad (8.1)$$

8.3 Marginal Static Stability, MaudNESS

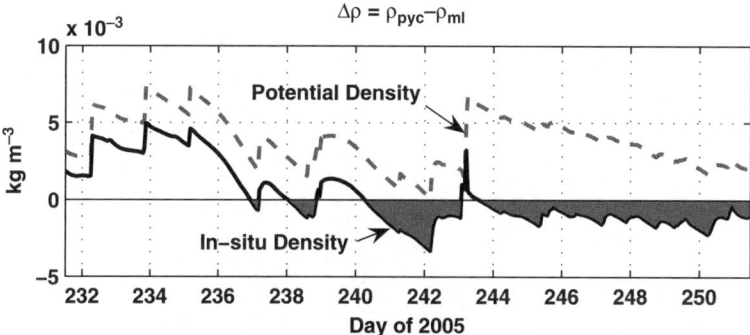

Fig. 8.22 Difference between density in the upper 5 m of the pycnocline and the well mixed-layer density, determined from potential density and density, where the latter refers to density calculated at pressure corresponding to the mixed layer/thermocline interface

In the potential density formulation, $\rho_i = \rho(T_i, S_i, p = 0)$. If instead, density in (8.1) were evaluated at pressures corresponding to zz_i and zz_{i-1}, the gradient (and buoyancy flux) would be dominated by fluid contraction with pressure, and would be useless for dynamical modeling. However, an approximation to the actual *in situ* density gradient is

$$(\rho_z)_i \doteq \frac{\rho_i[T_i, S_i, p(z_i)] - \rho_{i-1}[T_{i-1}, S_{i-1}, p(z_i)]}{\Delta zz_{i-1}} \tag{8.2}$$

i.e., where density is evaluated at pressure corresponding to the z_i grid point midway between the two zz grid points. For a sharp thermocline/halocline as in the model, this means that *in situ* N^2 is negative in the shaded portions of Fig. 8.22, thus a local instability exists.

Turbulence is enhanced in the LTC model whenever there is negative buoyancy flux within the scope of the dynamic boundary layer, i.e., within the zone influenced by surface stress and buoyancy flux conditions. Consequently, for the conditions identified above, mixing should be enhanced in a model that calculates density according to (8.2), which we designate the NES model, meaning some account is taken of nonlinearities in the equation of state. If surface conditions change—perhaps rapid melting markedly decreases turbulence scales in the upper part of the boundary layer—instabilities may persist below a level where surface driven turbulence is negligible. In that case an *ad hoc* assumption is that instabilities below the dynamic boundary layer (as defined above) will be rapidly relieved by other natural processes (e.g., internal waves). In practice, the NES model tests at each time step for negative N^2 in the region below the dynamic boundary layer, and averages those temperatures and salinities with adjacent values for which the gradient is stable. The NES model is identical to the PD model except for these two factors.

NES model temperature contours (Fig. 8.23) are similar to the PD model (Fig. 8.21, note that z scales are different) for the first half of the 20-day period. Starting on day 245, however, the NES model reaches static instability and rapidly

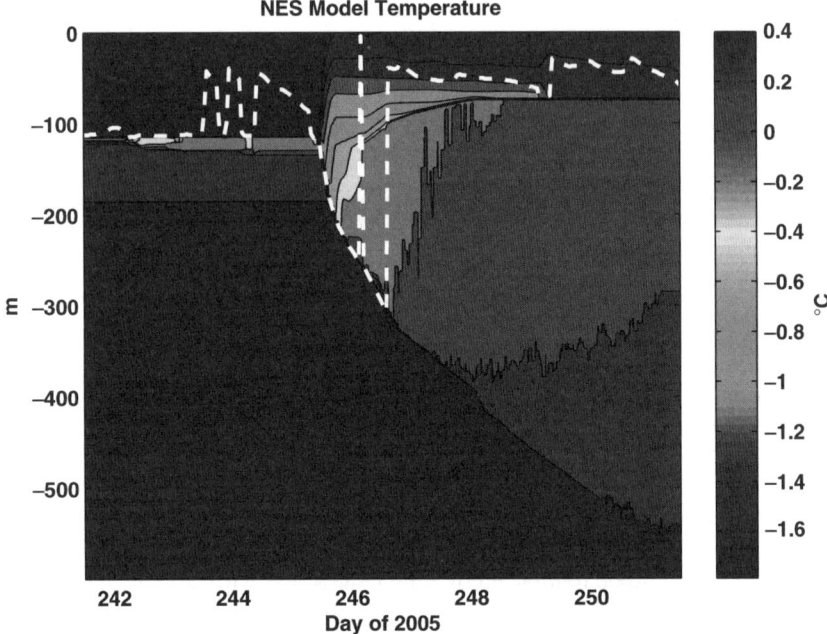

Fig. 8.23 NES model temperature contours in the last half of the simulation. The model becomes thermobarically unstable on day 245

mixes over a wide depth range. This has large impact on heat flux at the ice/water interface (Fig. 8.24), resulting in ice melt. During the initial breakthrough into the WDW layer, the dynamic boundary layer deepens rapidly (again shown as the white dashed curve in Fig. 8.23), except for a short time early on 246 when surface stress falls almost to zero, but then as the melting begins on day 246, the "thermal barrier" effect kicks in and limits the dynamic boundary layer to the upper 75 m or so. Despite the attenuation of surface stress by positive buoyancy at the interface, mixing continues at depth because of instabilities triggered by downward mixing of cold water from the upper layer, as demonstrated by Fig. 8.25. On day 246, the combination of melting and low surface stress begins formation of a new, shallower upper layer, yet a new, uniform property layer begins to form between about 100 and 200 m, and proceeds to grow both downward and upward. Layers like this are not uncommon in the upper ocean observed around Maud Rise, and this simple model illustrates how they might form.

The start and end temperature/salinity states for the NES model (Fig. 8.26) furnish one further observation. Despite a continuous loss of upper-ocean buoyancy over 20 days of heat loss to the atmosphere and about 4 cm of net ice growth, the thermocline has risen by about 25 m, and there is a reservoir of warm water nearer the surface than when the simulation started. Much of this re-arranging of the upper ocean properties, reaching depths more than 500 m, results from nonlinearities in the equation of state. Obviously, a simple one-dimensional model that

8.3 Marginal Static Stability, MaudNESS

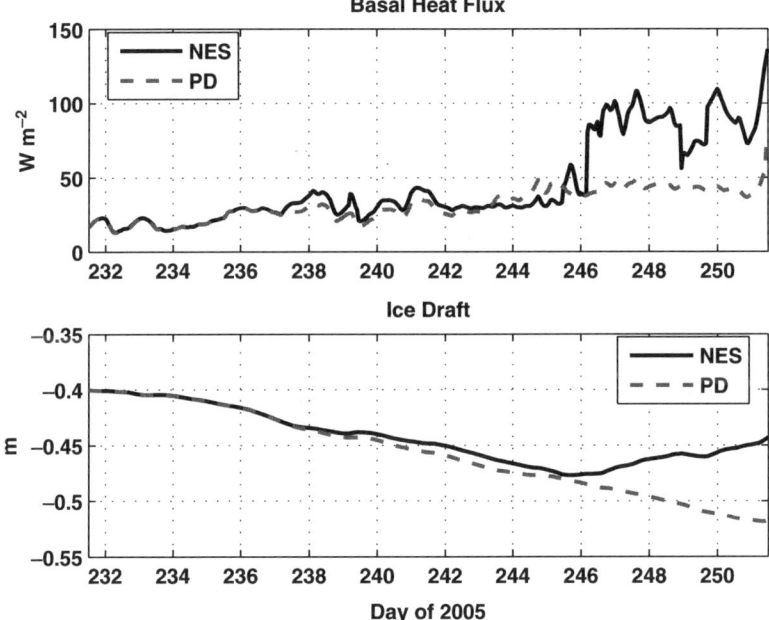

Fig. 8.24 Comparison of basal heat flux and ice draft for the two models with the same surface forcing

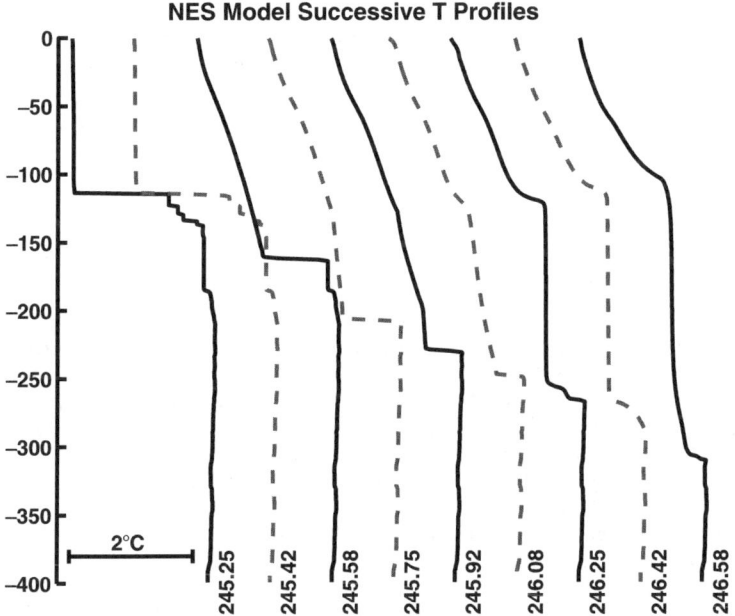

Fig. 8.25 Successive temperature profiles in the NES model during the time of intense downward mixing on days 245 and 246

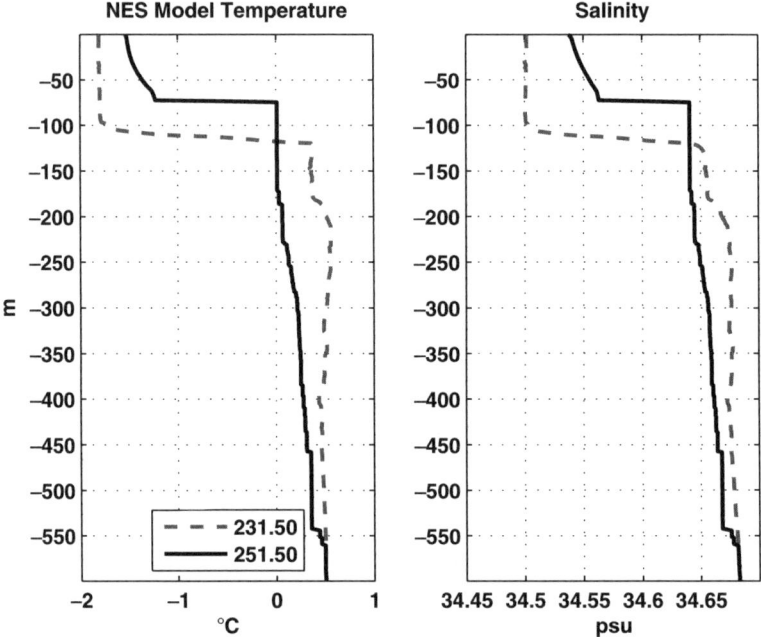

Fig. 8.26 Start and end temperature and salinity profiles for the NES model, run for 20 days with MM5 forcing, initialized to MaudNESS station 91

takes no account of horizontal gradients and associated geostrophic shear, as well as advection of WDW and Ekman transport in the IOBL, cannot expect to fully describe changes over three weeks. However, from an heuristic viewpoint, the results for the MaudNESS simulation suggest how some of the otherwise puzzling features encountered in the low stability regimes near Maud Rise come about.

References

Grenfell, T. C. and Maykut, G. A.: The optical properties of ice and snow in the Arctic Basin. J. Glac., 18, 445–463 (1977)

Light, B., Grenfell, T. C., and Perovich, D. K.: Transmission and absorption of solar radiation by Arctic sea ice during the melt season, J. Geophys. Res., 113, C03023, doi:10.1029/2006JC003977 (2008)

McPhee, M. G.: Analysis and prediction of short term ice drift. Trans. ASME J. Offshore Mech. Arctic Eng., 110, 94–100 (1988)

McPhee, M. G.: Marginal thermobaric stability in the ice-covered upper ocean over Maud Rise. J. Phys. Oceanogr., 30, 2710–2722 (2000)

McPhee, M. G.: Turbulent stress at the ice/ocean interface and bottom surface hydraulic roughness during the SHEBA drift. J. Geophys. Res., 107 (C10), 8037 (2002), doi: 10.1029/2000JC000633

McPhee, M. G. and Smith, J. D.: Measurements of the turbulent boundary layer under pack ice. J. Phys. Oceanogr., 6, 696–711 (1976)

References

Niiler, P. P. and Kraus, E. B.: One-dimensional Models of the Upper Ocean. In: Kraus, E. B. (ed.) Modelling and Prediction of the Upper Layers of the Ocean, pp. 143–172. Pergamon, Oxford (1977)

Perovich, D. K., Tucker, W. B., III, and Ligett, K. A.: Aerial observations of the evolution of ice surface conditions during summer. J. Geophys. Res., 107(C10), 8048 (2002), doi: 10.1029/2000JC000449

Pollard, R. T., Rhines, P. B., and Thompson, R. O. R. Y.: The deepening of the wind mixed layer. Geophys. Fluid Dyn., 3, 381–404 (1973)

Powers, J. G., Monaghan, A. J., Cayette, A. M., Bromwich, D. H., Kuo, Y. H., and Manning, K. W.: Real-time mesoscale modeling over Antarctica. Bull. Am. Meteorol. Soc., 84 (2003), doi: 10.1175/BAMS-84-11-1533

Chapter 9
The Steady Local Turbulence Closure Model

Abstract: A fundamental problem in boundary-layer physics is extrapolating limited measurements to a general description of the mean velocity and scalar properties, along with their Reynolds fluxes including values at the immediate boundary. For the atmospheric surface layer, extensive research has been devoted to methods relating relatively simple measurements to fluxes. Central to this approach is characterizing surface roughness for momentum and scalar variables. Typically, a tower is deployed with two or more levels of instrumentation and the surface fluxes are estimated either from the mean measurements across the tower using some form of the Monin-Obukhov dimensionless gradients (e.g., Businger et al. 1971; Andreas and Claffey 1995), or from a combination of mean gradients and fluxes, determined either by direct covariance or by spectral techniques (e.g., Edson et al. 1991).

In the IOBL, this is much less straightforward for a variety of reasons. First, in contrast to the upper sea-ice surface, variation in the underice morphology often occupies a significant fraction of the entire boundary layer. If the IOBL scales with about 1/30 of the atmospheric boundary layer, a pressure ridge with a 1-m sail and 5–6-m keel presents completely different aspects to the respective boundary layers. In general, for the IOBL parameterization problem, many of the surface-layer assumptions (constant stress, stress and mean velocity collinear with no direction change, etc.) are clearly inappropriate.

As illustrated in Chapter 8, it is sometimes possible to solve a time-dependent numerical PBL model with given initial conditions, letting it evolve in time as the forcing fields change. Given a suitable time series of observations at a particular location, to the extent that the model can reproduce the observed characteristics (say mixed layer temperature, salinity, depth), the model will provide a reasonably accurate description of the overall exchanges across the OBL. This depends on both having realistic initial conditions and a reasonably accurate time series of forcing fields (e.g., wind or ice velocity, conductive heat flux in the ice, etc.). In many cases, observations are scattered in both time and location (for example, stations taken from a ship or airplane during a regional survey), and one would like to produce a "snapshot" of the OBL structure, to estimate fluxes at the surface or near the base of the mixed layer.

Even when a relatively complete set of measurements exists, we are often faced with the sampling problem of extrapolating measurements made at a few location (which because of operational considerations, are often biased toward relatively smooth ice) to a general description for the entire surrounding ice field, which in turn might be appropriate to characterizing a grid cell in a numerical model (this is sometimes referred to as the "scaling up" problem). An example from ISPOL (McPhee 2008, in press) serves to clarify this problem. The floe with which we drifted north in the western Weddell Sea comprised a conglomerate of several different ice types including heavily ridged portions, relatively thin ($\sim$1 m) regions of first year ice, plus reasonably smooth regions of multiyear ice about 2 m thick. For most of the project the turbulence mast was located under ice of the last type, with the undersurface in the immediate vicinity quite smooth, but with pressure ridges and the floe edge within the first 100 m or so from the site. Toward the end of the drift phase of the project (on December 25) the ice floe split, forcing relocation of the turbulence mast, which for the last week of the project was located under thin ice near a small pressure ridge.

During the first deployment, we consistently observed a substantial increase in turbulent stress with depth across the 6 m span of the turbulence mast (see Fig. 9.10 of McPhee 2008), which we interpreted as the deeper sensors picking up turbulence generated by large undersurface features some distance away. This phenomenon has often been observed in other projects as well, typically where the mast was located under smooth ice, but there were roughness features within a distance given roughly by the ratio of mean velocity to scale velocity (u/u_*) times the depth of the turbulence sensor measured from the interface (Morison and McPhee 2001). So, for example, a TIC 2 m below the boundary might sense roughness features within about 30 m, whereas turbulence measured 4 m lower might respond to undersurface protrusions up to 100 m away. This rule of thumb seemed to hold reasonably well for SHEBA as well as ISPOL (McPhee 2002).

For the short deployment at the end of the project, a mast with two clusters, 1 and 3 m below the ice undersurface, respectively, was initially placed so that the predominant tidal flow would approach from the north or south across relatively smooth ice, and parallel to a small pressure ridge situated to the west. Soon after deployment, however, the floe rotated so that if the current sensed by the mast came from the northeast, the keel was directly upstream from the turbulence mast, with large impact on flow in the upper few meters. The installation included an acoustic Doppler profiler that provided high-resolution current profiles from about 10 to 30 m depth. Two examples from this second installation are shown in Fig. 9.1: one with flow approaching the mast across smooth ice, and the second with relative flow almost directly across the pressure ridge keel. In the former case, the current structure shows a reasonably well developed Ekman spiral, with friction velocity at 1 m of about 5 mm s^{-1}, and from direct application of the LOW to the dimensionless velocity, we infer a surface roughness of about 0.8 mm. When flow approaches from across the keel, the hodograph from ADP data in the range from 10 to 30 m again exhibits the expected Ekman turning, but now currents at the TIC depths are a small fraction of the deeper currents, indicating flow blockage. If u_* estimated from the

9 The Steady Local Turbulence Closure Model

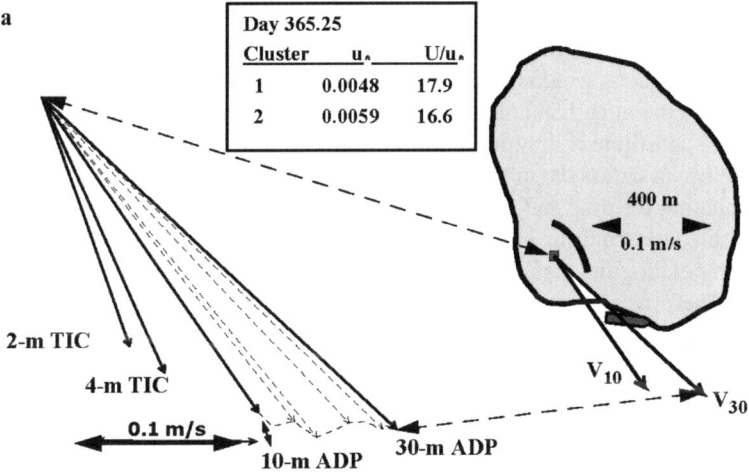

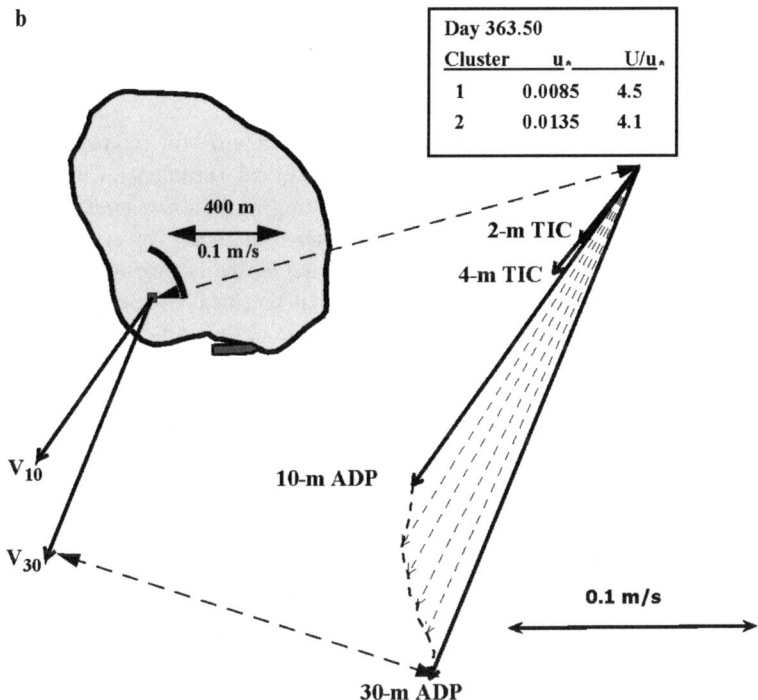

Fig. 9.1 Currents measured relative to the drifting floe during the second deployment at ISPOL. The floe outline and orientation are shown, with mast location indicated by the square symbol and ADP current vectors at 10 m and 30 m depth. North is up. A more complete current hodograph with the TIC currents is shown in blowup view. Boxes list the friction velocity and dimensionless current. The heavy black curve represents the position of a small pressure ridge. **a** 3-h average centered at time 365.25 (31 January 2004) with flow approaching from smooth ice, and **b** at time 363.5 with flow from across the ridge keel

covariance statistics at 1-m is taken at face value, z_0 is about 17 cm. With that much flow disturbance from the pressure ridge, many of the assumptions underlying the flux determinations would be suspect, and we would normally flag such data as unreliable. On the other hand, the example illustrates that form drag on pressure ridge keels will constitute a significant part of the total momentum transfer between the floe and ocean unless the undersurface is exceptionally smooth.

Evaluating the drag and enhanced mixing from even one pressure ridge keel is a formidable task requiring extensive computation (see, e.g., Skyllingstad et al. 2003), and extrapolating the results to an entire heterogeneous floe adds considerable difficulty. There is, however, a hint in the deeper current profiles in Fig. 9.1 that by considering what happens in the outer part of the IOBL, it may be feasible to infer surface properties representative of the entire floe, the main point being that because the floe moves as a rigid body, at depths greater than most of the undersurface protrusions, the turbulence must sense some integrated impact the varying surface conditions. In this chapter, we explore this concept with a modeling technique developed from the ISPOL measurements (McPhee 2008, in press).

9.1 Model Description

Unlike scalar conservation equations, the Ekman equation for momentum admits a steady-state solution. A "steady" version of the Local Turbulence Closure model (SLTC) was developed as a means of extrapolating limited measurements at particular times to deduce the structure of the entire boundary layer. The primary assumption and simplification for the SLTC model is that turbulence adjusts in effect instantaneously to surface conditions so that the local time-dependent terms in the conservation equations are negligible relative to the vertical exchange terms (e.g., for momentum $|u_t| \ll |\tau_z - if u|$) and that the vertical transport of TKE is not a major factor in most IOBL instantiations. While these assumptions are suspect when large inertial oscillation is present, or during rapid changes in surface flux conditions, they nevertheless often persist for reasonably long periods, especially when the ice cover is compact. In practice the model requires a reasonably good description of the temperature and salinity structure of the upper ocean, and some way of estimating friction velocity at the interface, perhaps from ice velocity or surface wind (if ice is drifting freely). As explained below, the model utilizes an iterative scheme that first estimates the IOBL eddy viscosity solely from surface flux conditions. Then by Reynolds analogy, it estimates scalar fluxes using eddy diffusivity based on the modeled eddy viscosity. In general, these fluxes will affect the turbulence scales and eddy viscosity, so the steady momentum is solved again with the new eddy viscosity, fluxes are re-calculated, and so on. We demonstrated (McPhee 1999) that using this model to simulate the time evolution of temperature and salinity in the upper ocean produced results similar to a simulation using a second-moment closure model (level $2^1/_2$ of Mellor and Yamada 1982). The latter required forward stepping of six conservation equations, while the SLTC time stepped only the T and S fields. In terms of computing cost, there was no great advantage, since the iterative scheme is computationally expensive, but the point was to show that the purely local (in space *and* time) turbulence description produced similar fluxes as the more sophisticated model which carried additional equations for momentum, TKE, and master length scale.

The model employs essentially the same physics as the time-dependent model described in Chapter 7, except that rather than stepping forward in time from an initial state, forced by prescribed surface conditions, it considers a fixed upper ocean temperature and salinity state, with one set of interface flux conditions and iterates to a solution for momentum and scalar fluxes based on a physically reasonable distribution of eddy viscosity and scalar diffusivity.

9.2 The Eddy Viscosity/Diffusivity Iteration

Unlike the time-dependent model, where for each time step the buoyancy flux and eddy diffusivities are determined from a previous time step, the "stand-alone" SLTC model begins from an initial guess at buoyancy flux, then iterates to a solution in which the modeled u_* and observed T/S profiles determine the boundary-layer structure.

To illustrate the method, consider 3-h average profiles of potential temperature and salinity from late in the SHEBA project (Fig. 9.2) and assume that $u_{*0} = 18$ mm s^{-1} is prescribed. This is used along with T and S in the upper ocean to calculate $\langle w' b' \rangle_0$. An initial guess for eddy viscosity (Fig. 9.3a) is made by determining a maximum value $K_{\max} = u_{*0} \lambda_{\max}$ where $\lambda_{\max}$ is determined from u_{*0} and $\langle w' b' \rangle_0$ according to the algorithm described in Section 7.6. An exponential falloff in stress is assumed

$$K_m \text{ (initial)} = u_{*0} \lambda_{\max} e^{\sqrt{\frac{|f|}{2 K_{\max}}} \cdot \hat{z}} \tag{9.1}$$

except in the surface layer where it varies as $\kappa |z| u_{*0}$. This estimate assumes neutral stability throughout the water column, so that scalar diffusivity equals viscosity, which remains unrealistically large far past the mixed layer depth (indicated by the dashed line in Fig. 9.2b). As the arrow from a to b indicates, applying scalar diffusivity to the observed θ and S profiles provides an initial estimate of buoyancy flux through the entire OBL, which is also unrealistically large below the mixed layer. By applying the mixing length algorithm with the first model estimate of profiles for $\langle w' b' \rangle$ (Fig. 9.3b) and u_{*0} along with specified interface fluxes, a second K_m estimate follows (Fig. 9.3c), from which new estimates are made (Fig. 9.3d), and so on, for a specified number of iterations. Results for eddy viscosity and buoyancy flux after the next iteration are shown in Fig. 9.3e and f, along with results after ten iterations (gray curves).

Details of the simulated eddy viscosity are shown in Fig. 9.4, along with estimates of eddy viscosity from two TICs, calculated from the products of local friction velocity and mixing length (inversely proportional to the wave number at

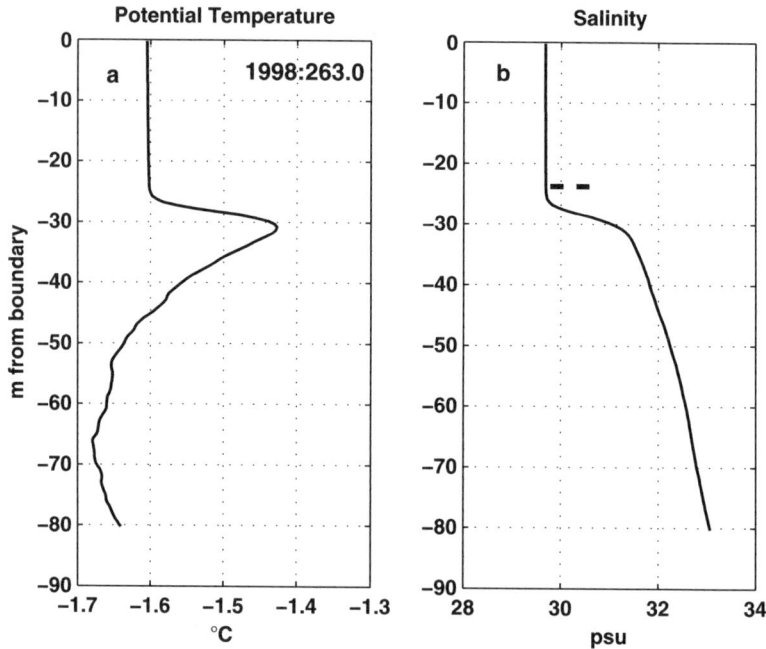

Fig. 9.2 SHEBA profiler potential temperature and salinity from 3-h average centered at 00:00UT on 20 September 1998, used to illustrate the SLTC model. The dashed line indicates the last grid point in the well mixed layer

the maximum in the w spectrum). In the upper part of the pycnocline (Fig. 9.4b), K_m decreases exponentially with distance measured from the pycnocline depth, defined as the level at which squared buoyancy frequency first exceeds a minimum level, in this case $2.5 \times 10^{-5}\,\text{s}^{-2}$. A combination of u_{*p} and $\langle w'b' \rangle_p$ from the model solution at the pycnocline level determines λ in the upper pycnocline. In the stable stratification of the pycnocline, the ratio of scalar diffusivity to eddy viscosity is a function of gradient Richardson number (Section 7.6) since turbulence is more effective at transferring momentum than scalar properties. This leads to a more rapid decrease with depth in K_h.

Modeled Reynolds stress, from the product of K_m and the numerical velocity gradient, is shown as u_* (the square root of kinematic stress magnitude) in Fig. 9.5a, along with the measurements. For this demonstration, u_{*0} was chosen so that the modeled stress matched measured at the lower instrument cluster (6 m below the ice) by successive adjustments to an initial guess assuming an exponential falloff from the interface to the 6-m level. Modeled heat flux $(-\rho c_p K_h \theta_z)$, shown in Fig. 9.5b, indicates an upward flux of roughly $10\,\text{W}\,\text{m}^{-2}$ in the upper part of the well mixed layer, in agreement with measurements $(\rho c_p \langle w'T' \rangle)$ at the instrument cluster levels. Note that the only "modeled" part of the heat flux profile is the eddy diffusivity, K_h. The results are consistent with the interface flux dependent on the elevation of mixed layer temperature above freezing (square symbol). In the lower part of the

9.2 The Eddy Viscosity/Diffusivity Iteration

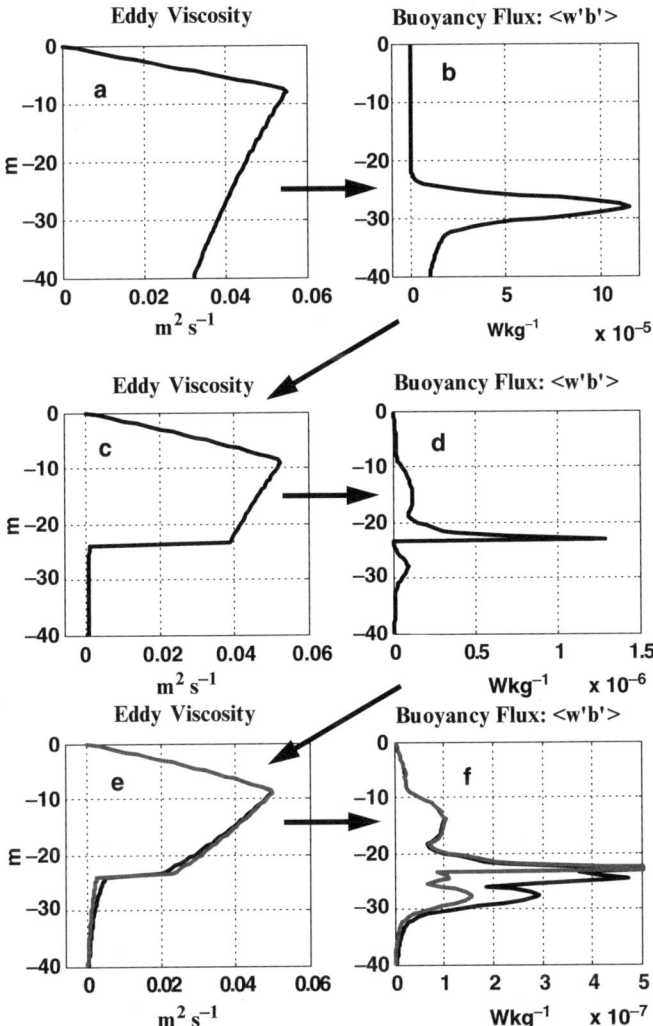

Fig. 9.3 a Initial eddy viscosity (K_m) guess based on exponential u_* profile and λ calculated from surface fluxes. **b** Buoyancy flux profile from initial guess. **c** First iteration revised K_m profile including buoyancy flux from *b*. **d** Second buoyancy flux estimate. **e** Second (black) and last (gray) eddy viscosity iterations. **f** Third (black) and last (gray) buoyancy flux estimates. Note the scale changes in buoyancy flux estimates

well mixed layer, the modeled heat flux is about twice as large indicating active mixing of heat from below. Presumably the flux divergence would heat the mixed layer as time progresses. This is, of course, an instantaneous snapshot, but indicates how the "steady" model may be used to estimate temporal evolution of upper ocean scalar properties (McPhee 1999).

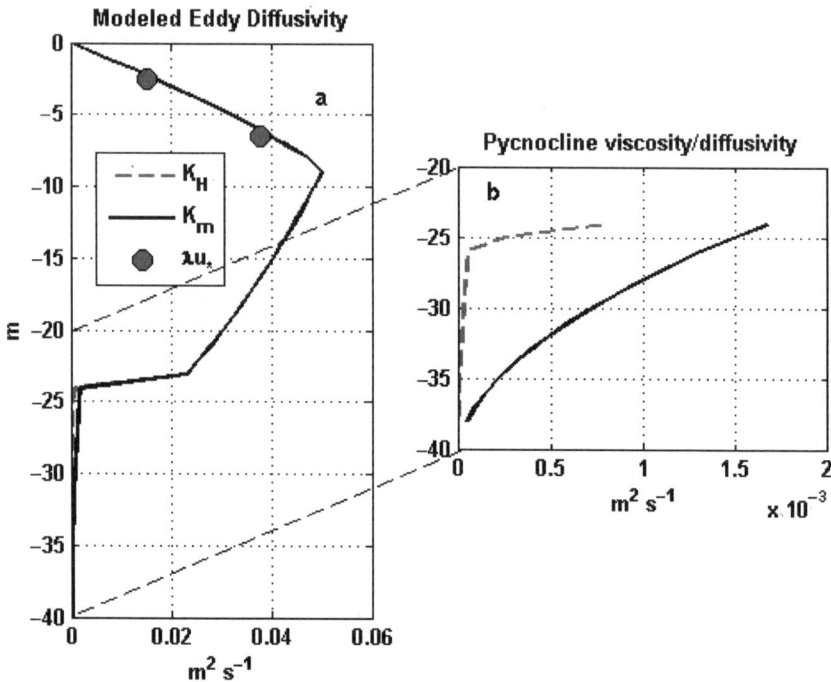

Fig. 9.4 Eddy viscosity and thermal diffusivity after the iteration of Fig. 9.3, for the upper 40 m of the 80 m model domain **a** and detail in the pycnocline showing the reduction of scalar diffusivity relative to viscosity **b**

In the upper 10 m or so of the pycnocline (beginning at about 24 m) there is still relatively strong mixing of both momentum and heat, despite the rapid attenuation of eddy diffusivities because of upward buoyancy flux. Note that heat flux falls off in the pycnocline at about the same rate as momentum flux, even though the eddy thermal diffusivity is much smaller than eddy viscosity.

To recap, the demonstration shows that given measured profiles of T and S encompassing the well mixed layer and pycnocline, along with Reynolds stress measured at one level, a plausible distribution of momentum and scalar fluxes throughout the entire boundary layer may be constructed, including estimates of the interfacial fluxes. More information is required, however, to characterize the entire velocity structure (with respect to the undisturbed ocean velocity), namely, the undersurface hydraulic roughness, z_0. Generally, pack ice measurements are made from a platform that is moving relative to the underlying undisturbed ocean, and water velocity measured from the ice is not the absolute velocity in a fixed-to-earth reference frame, but rather the vector difference between the absolute velocity at the measurement depth and the ice velocity. With modern satellite navigation, the latter may be measured quite accurately, and provided the orientation of the instruments is known (not always a trivial problem when dependent on compasses at high latitudes), it is a simple matter to determine the absolute velocity, say for example,

9.2 The Eddy Viscosity/Diffusivity Iteration

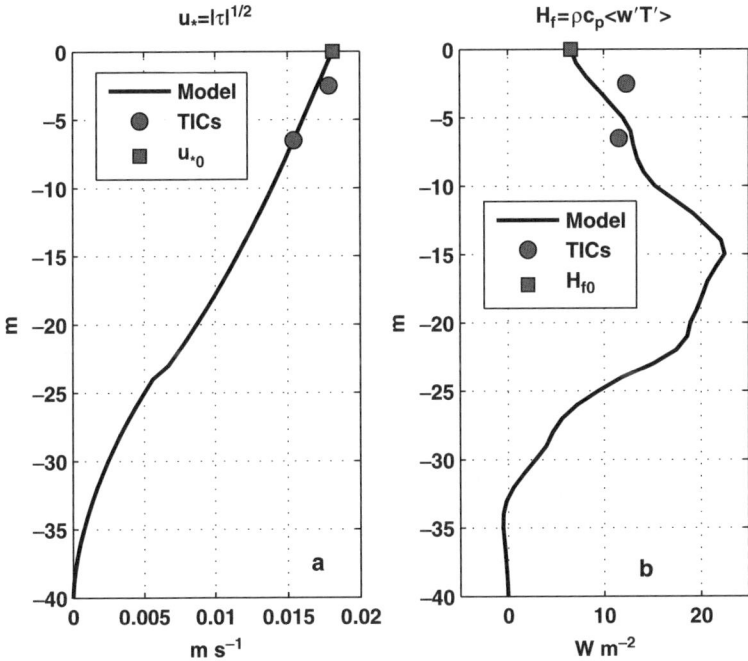

Fig. 9.5 a Friction velocity (square root of kinematic Reynolds stress) as modeled (solid) and measured at two levels. The surface value (square) symbol was chosen so that the modeled value at 6 m matched measured. **b**. Corresponding profiles and measurements of turbulent heat flux. The interface value (square) is calculated from u_{*0}, T and S in the upper ocean

6 m below the interface. But in a well developed, turbulent boundary layer, the 6-m current comprises contributions from stress-driven shear in OBL, any inertial motion in the phase-locked ice/upper ocean system, plus the geostrophic current arising from slope in the sea surface. The last is the current that would exist without any shear between the ice and undisturbed ocean.

Since there is no provision in the SLTC model for inertial oscillations and in general the vector sum of geostrophic and inertial velocities at the measurement level is unknown, the surface roughness may be estimated from current measurement at a particular level as diagrammed in Fig. 9.6. The premise is that the topmost point in the mean quantity (zz) grid is within the surface layer so that surface stress and shear are aligned, in which case the velocity difference between the topmost grid point and the ice obeys the law of the wall:

$$\frac{\kappa \Delta u}{u_{*0}} = \ln \frac{|zz_1|}{z_0} \quad (9.2)$$

Geometrically, Δu is determined by the intersection of an arc with length equal to the magnitude of the measured current (indicated by $|V_{m(rel)}|$), swung from the tip of the absolute model velocity vector ($V_{m(abs)}$) at the measurement depth, and a line extended in the direction of u_{*0} from the velocity at the topmost grid point.

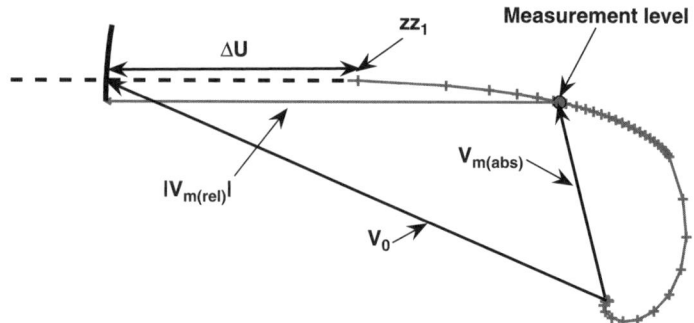

Fig. 9.6 Diagram showing how measurement of velocity at one level in the IOBL may be used to estimate the surface velocity and undersurface hydraulic roughness

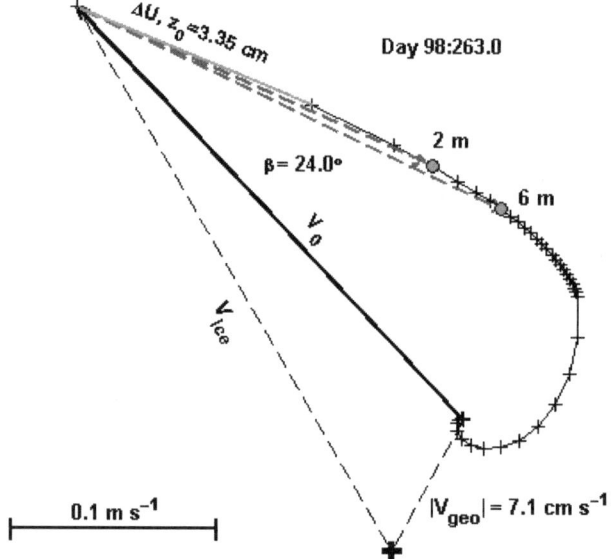

Fig. 9.7 The model hodograph from the September 20 example. Velocity measured relative to the ice at 6 m is used to scale the surface layer velocity, hence V_0. The ice velocity from satellite navigation, V_{ice}, includes inertial and geostrophic shear effects not modeled. North is up

A vector extending from the model coordinate origin to this point then represents the ice velocity (V_0) relative to the underlying ocean in the model reference frame. The model is then oriented by aligning the modeled and measured relative current vectors at the measurement level. This entails multiplying horizontal velocity vectors by a complex factor:

$$r = \frac{V_{obs}}{V_{m\,(abs)} - V_0} \tag{9.3}$$

The entire velocity solution for the example is diagrammed in plan view in Fig. 9.7. The model velocity relative to an observer on the ice matches the observed current

9.2 The Eddy Viscosity/Diffusivity Iteration

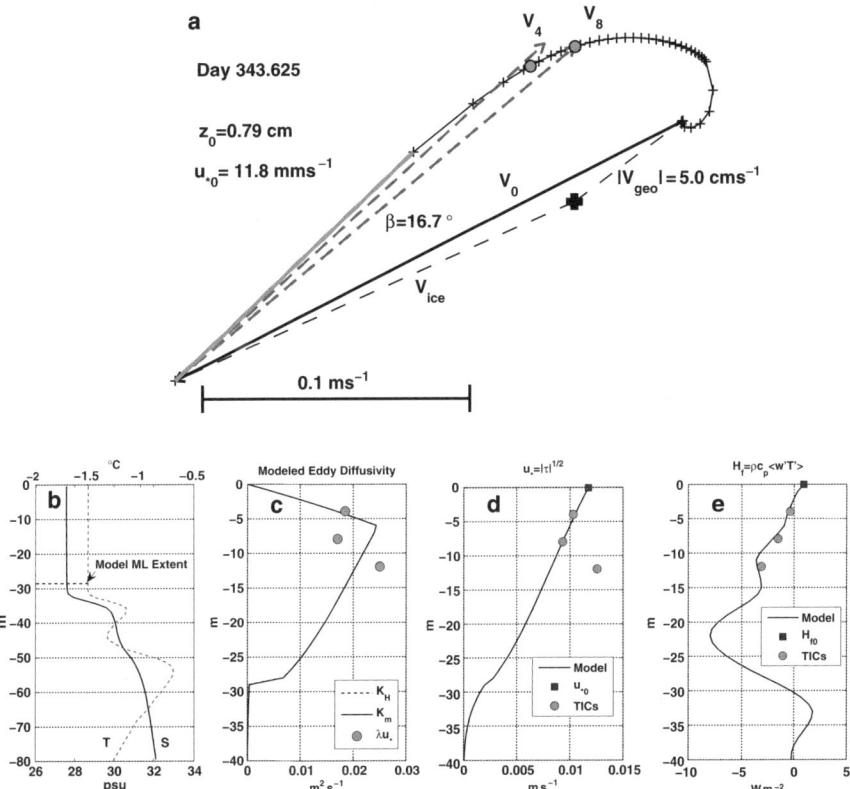

Fig. 9.8 SLTC model realization for 3-h periods centered at 15:00 UT on 9 December 1997 during SHEBA. **a** model hodograph; **b** observed potential temperature and salinity profiles in model domain; **c** eddy viscosity/diffusivity. **d** friction velocity; and **e** turbulent heat flux

at 6 m, and establishes V_0, the vector ice velocity relative to the undisturbed ocean. This differs from the actual ice velocity obtained by satellite navigation (V_{ice}) by "V_{geo}" where quotes indicate that this is a combination of actual geostrophic flow plus any inertial or baroclinic motions, which are not considered in the SLTC model.[1]

One other example from early in the SHEBA project (Fig. 9.8) demonstrates a rare period during winter when there was downward turbulent heat flux in the water column, despite a lack of short wave radiation (the sun had set) and enough ΔT to imply a positive basal heat flux of about $1\,\mathrm{W\,m^{-2}}$. There was a rather dramatic increase in stress from 8 to 12 m (Fig. 9.8d), probably from enhanced stirring by a pressure ridge keel about 110 m to the SW. Although not seen at the scale shown in Fig. 9.8b, there is a positive potential temperature gradient in the

---

[1] "V_{geo}" will also reflect any uncertainty in alignment of the turbulence mast, which often depends on compass headings and a model for magnetic declination.

well mixed layer that combined with the calculated eddy thermal diffusivity to produce the downward heat flux in the upper ocean, in agreement with the observations. A possible explanation for the positive temperature gradient is that all during the day of 9 December (day 343), as the ice drifted to the southwest, temperature of the well mixed layer decreased steadily by a total of about 12 mK (from -1.493 to $-1.505\,°C$). From Fig. 9.8a, the *absolute* velocity hodograph shows that water in the upper part of the column was transported from NW (warmer) to SW (cooler) faster than in the lower column, resulting in a modest downward heat flux.

9.3 Applications

9.3.1 Ice Station Polarstern

The ISPOL project in the multiyear ice pack of the western Weddell Sea examined early summer air-ice-ocean interaction from a wide range of physical, biological, and gas-exchange perspectives (Hellmer et al. 2008, in press). A central issue was characterizing the turbulent exchange of scalar contaminants (including nutrients and biota) between the upper ocean and the sea ice during a time of maximum solar radiation. This presented a challenging problem in an ice pack with large variation in ice thickness, evidence of extensive ice deformation, and several embedded icebergs drifting with the sea ice within sight of the ship. Adding to the complexity was that, in contrast to summer pack ice in the central Arctic, the main driving force was not wind, but rather a combination of tidal and baroclinic currents just offshore of the eastern Antarctic Peninsula continental shelf.

As Fig. 9.1 suggests, applying measurements from a relatively smooth site to the entire floe may miss important parts of the momentum transfer as well as heat and salt exchange. Our ISPOL observations indicated that turbulent stress consistently increased with measurement depth, which meant that the effective surface roughness, z_0, also increased with distance from the boundary (Fig. 9.9). There is rough agreement between z_0 estimates from the LOW and from a modification that considers "measured" mixing length (McPhee 2002), except for the shallow cluster at OT-II, which was often affected strongly by flow blockage. The question posed is: which, if any, measurement level is representative of the entire floe? Currents measured by the acoustic Doppler profiler in the range from 10 to 30 m at both sites almost always showed counterclockwise (Ekman) deflection with increasing depth (in the drifting reference frame). We reasoned that these levels were below most of the obstructions on the ice underside, and that if currents were averaged over many directions and different current speeds, the resulting average current would reflect the integrated impact of varying undersurface morphology.

Of the 3-h averages of ADP current profiles between 10 and 30 m depth, there were 82 profiles (with acceptable signal-to-noise ratios) where current speed at 30 m

9.3 Applications

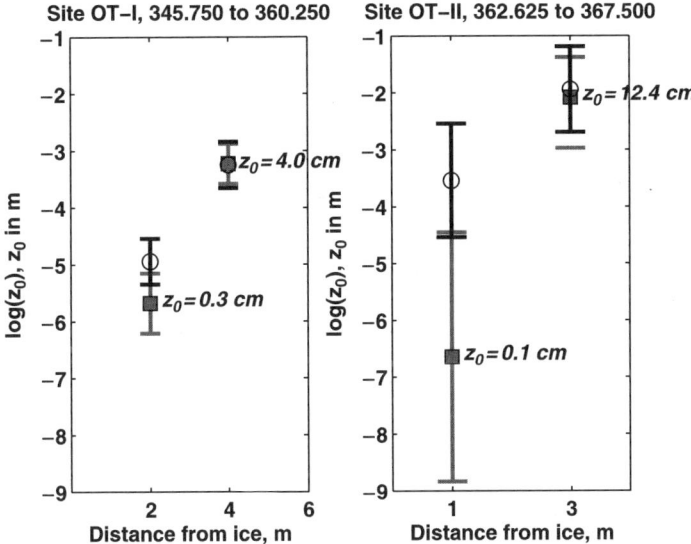

Fig. 9.9 Median values of $\log(z_0)$ for the main ISPOL turbulence mast site in December 2004. Squares are from a mixing length method (McPhee 2002); circles are application of the LOW. Error bars are 95% confidence limits (Adapted from McPhee 2008)

was greater than or equal to 0.06 m s^{-1}. Each of these profiles was nondimensionalized by dividing the complex (vector) current at 2 m sampling intervals by the complex current at 30 m. The dimensionless current hodograph then produced a smooth spiral shape with about 15° of counterclockwise rotation as depth increased from 10 to 30 m (McPhee 2008, in press).

We reasoned that an estimate of overall surface roughness that was independent of the turbulence measurements and perhaps free from local topographic effects could be made as follows. First, specify a trial value for z_0. Next, for each acceptable current profile, use the SLTC model to calculate the current profile. This was done by (i) specifying upper ocean T and S profiles, by interpolating in time from twice daily ship CTD stations; and (ii) forcing the model to match the measured current at 20 m, chosen to be generally in the well mixed layer, but deep enough to be away from the immediate impact of ridge keels. Since the object is to avoid using TIC data, for any particular model run, u_{*0} is first estimated from a Rossby-similarity calculation, then adjusted iteratively until the model (relative to ice) velocity matches the ADP velocity at 20 m. We did this for three different values of z_0, over the range of median values shown in Fig. 9.9. For each 3-h model run (of 43 total), the modeled currents were nondimensionalized by the model 30 m current (in a reference frame attached to the ice), then averaged. Results (Fig. 9.10) indicate that turning between 10 and 30 m was best modeled with z_0 equal to approximately 4 cm, the value found for 4 m TIC at the main OT-I site. This not only confirmed that the floe was relatively rough, but also provided a way of estimating the total floe-average ocean/ice fluxes of momentum, heat, and salt during the entire ISPOL project.

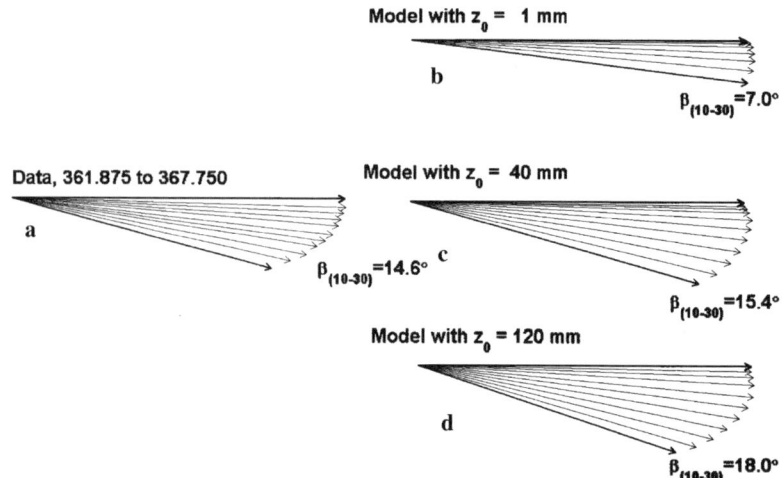

Fig. 9.10 a. Average of 43, 3-h average current hodographs divided by the current vector at 30 m (horizontal vector) for times when $|V_{30}| \geq 0.06\,\text{m s}^{-1}$. Vectors are drawn every 2 m from 10 to 30 m. The total Ekman angular shear between 10 and 30 m is $\beta_{10-30} = 14.6°$. **b, c** Average model dimensionless hodographs for the same times with three different z_0 values (From McPhee 2008)

9.3.2 Underice Hydraulic Roughness for SHEBA

A primary aim of the year-long SHEBA project was to characterize multiyear ice in the western Arctic at scales useful for large scale modeling of air-ice-ocean interaction, with particular attention to important terms in the surface energy budget. As discussed earlier, we often observed during the year-long project that turbulence increased with increasing distance from the ice/ocean boundary. This suggested that deeper clusters were sensing upstream obstacles at increasingly distant fetch, including a prominent pressure ridge roughly 100 m away. We noted large variations in apparent roughness as drift direction and floe orientation varied over the year-long deployment. By considering, stress, velocity, and apparent eddy viscosity at the TIC nearest the interface, we developed a technique that included λ as inferred from spectral peaks as an independent parameter (McPhee 2002). This minimized the effect of upstream heterogeneity in the flow for determining the local hydraulic roughness of undeformed ice, and provided an estimate of around 6 mm for the undersurface hydraulic roughness. We emphasized that this value was *not* indicative of the "aggregate" floe roughness, which would include added drag from ridge keels and floe edges, or reduction from open water or smooth ice.

During the relocation of the SHEBA oceanography program after the floe breakup in March 1998, we drilled undeformed ice in numerous location, looking for slightly thicker hummocks from the previous summer melt period, obscured at the surface by the winter snow accumulation. These were thought to be the most likely locations for siting instruments and shelters that would survive the upcoming summer melt. We typically found about 20 cm difference between ice thickness in hummocks versus "fossil" melt ponds. Laboratory studies show that hydraulic roughness is typically about 1/30 the "grain size" of the elements contributing roughness. For the portions of the SHEBA floe away from pressure ridges and leads, $z_0 = 6$ mm would imply a "grain size" of about 20 cm, hence is not inconsistent with our limited observations.

The question remains: what is the "aggregate" roughness of a typical multi-year Arctic ice floe in the region traversed by SHEBA? The previous analysis (McPhee 2002) utilized the cluster nearest the interface (nominally 4 m until summer, when it was raised to 2 m), and used an estimate of the mixing length there to adjust shear between the interface and measurement level, which tended to decrease z_0 from its LOW value at SHEBA site 1 (November to mid-March), and increase it slightly at site 2 (mid-March through September) relative to the LOW estimate. Measurements at the former site were obviously affected by a pressure ridge keel that was often "upstream" as the station drifted west and later north, while site 2 was farther from any apparent features. When averaged over the entire site 1 deployment, there was a monotonic increase in average u_* with depth from clusters 1 to 3 (nominally 4–12 m from the ice).

To address the SHEBA "scaling up" problem, we speculated that a technique analogous to that developed to characterize the ISPOL floe could be adapted for the SHEBA data. The approach settled on was somewhat different. Although there was an acoustic Doppler profiler at SHEBA, its data return rate over the course of the deployment was disappointing, and there often appeared to be spurious returns in the upper portion of the current profiles that contaminated measurements within the well mixed layer. We chose instead to use Reynolds stress and current measurements from TIC 2 on the turbulence mast, along with T/S profiles from the SHEBA automated profiler to solve the SLTC model for each 3-h average in the period from 15 November 1997 to 1 June 1998 when ice drift speed (without the inertial component) exceeded 0.1 m s^{-1}. Cluster 2 was chosen because it was at a depth (nominally 8 m from the ice) thought to minimize the impact of upstream heterogeneity and because it had the most samples (clusters 3 and 4 were sometimes below the well mixed layer, and were not redeployed after the March 1998 breakup).

A time series of $\log(z_0)$ for each of 249, 3-h model realizations meeting the minimum velocity requirements (also excluding about 24 samples where the derived z_0 was smaller than 6 mm) is shown in Fig. 9.11, along with averages in ten-day bins. The mean value with standard deviation error bars is $\log(z_0) = -3.0 \pm 1.0$. The mean value is thus about 4.9 cm with a range implied by the standard deviation of $\log(z_0)$, $1.6 \leq z_0 \leq 14.6$ cm.

The dimensionless surface velocity $\Gamma = V_0/u_{*0}$, shown in Fig. 9.12 for each model realization, is the complex inverse of a "geostrophic drag coefficient" that includes turning angle as well as magnitude. In terms of a more conventional quadratic drag coefficient, c_w (where $\tau_0 = u_{*0}^2 = c_w V_0^2$), the mean magnitude of Γ implies $c_w = 0.0056$. This is nearly equal to the value of 0.0055 derived in a different manner from the free-drift force balance for the AIDJEX stations in the central Beaufort Gyre in 1975 (McPhee 1980). The latter depended on a relatively high value for the 10-m wind drag coefficient based on analysis of balloon soundings during AIDJEX (Leavitt 1980; Carsey 1980). The mean turning angle inferred from the AIDJEX free-drift force balance was slightly less: 23°.

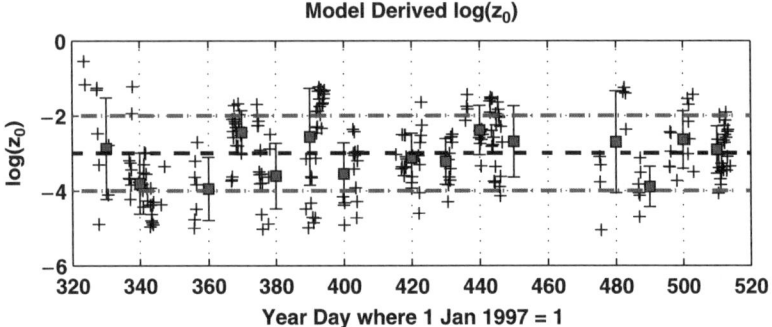

Fig. 9.11 Time series of the natural logarithm of underice surface roughness derived from SLTC model runs based on measured Reynolds stress and relative velocity at TIC 2 on the SHEBA turbulence mast, nominally 8 m below the ice/ocean interface (see also Plate 27 in the Colour Plate Section)

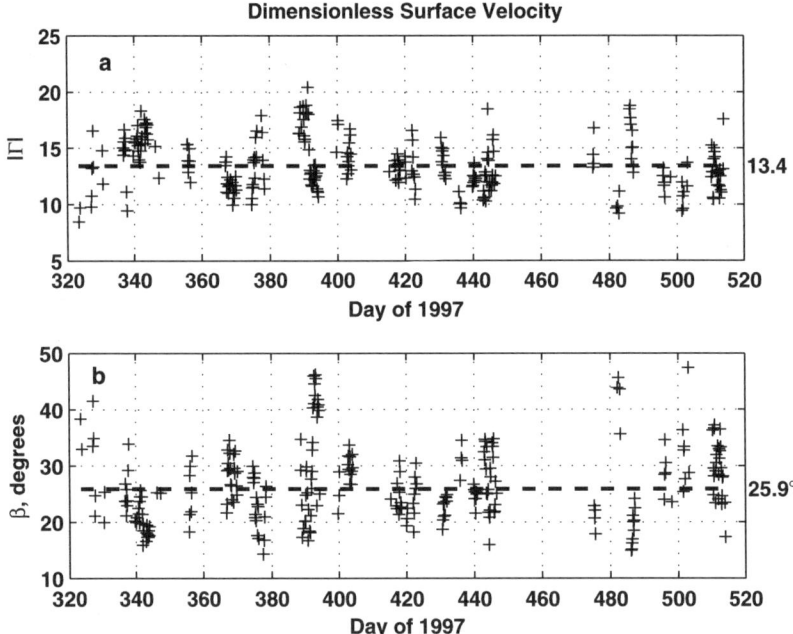

Fig. 9.12 a Magnitude of dimensionless surface velocity (with respect to undisturbed flow) for the model solutions. **b** Magnitude of the angle between V_0 and u_{*0}

In contrast to the AIDJEX analysis, the magnitudes of Γ and β do not decrease with increasing V_0 as would be expected if the boundary layer strictly followed Rossby similarity scaling. Indeed, the trend is opposite, with Γ increasing in both magnitude and deflection angle. The most plausible explanation is that during almost all of SHEBA the well mixed layer was significantly shallower then during corresponding time at AIDJEX. In similarity terms, the nondimensional pycnocline depth was thus particularly small for higher speeds during SHEBA, so that momentum flux was more confined vertically which tends to increase both the dimensionless surface velocity and the amount of OBL turning.

9.3.3 SHEBA Time Series

A primary application of the SLTC is estimating time series of IOBL characteristics from limited data. During SHEBA, the geometry of the TIC mechanical current meter triads limited the effective threshold velocity to about 0.05 m s^{-1}. In the summer of 1998, biological activity in the upper ocean often also degraded the turbulence data by fouling the current meters. Thus averages estimated only from times when the TICs operated will most likely be biased high.

SHEBA provided more or less continuous records of upper ocean temperature and salinity from about 1 November 1997 until late September 1998, with bursts of several continuous profiles at least twice per day. There were a few gaps of several days for instrument repair and during breakup events. There was also a complete record of GPS ship positions, providing accurate ice velocities. Under the assumptions that (i) the mean value for z_0 found above is representative of the entire floe for the duration of the project, and that (ii) over the long drift the average value of geostrophic (sea-surface tilt) current is near zero, a continuous time series of friction velocity and heat flux at the ice ocean interface was obtained by solving the SLTC model every 3 h, using ice velocity (after removal of inertial, and possibly tidal components by complex demodulation); temperature and salinity interpolated linearly between adjacent 3-h averages for which there were at least two up and down profiles; and estimates of temperature gradient in the lower ice column from the thermistor records of ice observation station "Pittsburgh" (Perovich et al. 1999). Results of the calculations, divided into three-month quarters, are shown in Fig. 9.13. In early winter (Fig. 9.13a), the weather was relatively stormy, with several events in which friction velocity exceeded 1.5cm s^{-1}. However, despite a relatively shallow mean pycnocline depth of about 17.5 m, basal heat flux (mean 0.3W m^{-2}) was remarkably small, underscoring the impact of the strong salinity gradient across the pycnocline. After the station drifted away from the anomalously fresh surface water toward the end of January 1998, there were several storm events in which basal heat flux exceeded 20W m^{-2}. The average remained relatively small (Fig. 9.13b), but was still an order of magnitude greater than the previous two months, while mean stress was almost the same. Most of the heat flux in this quarter occurred in late March. The SHEBA profiler was without power thus inactive for nearly four days starting on 17 March 1998. Because the model interpolates T/S profiles, it therefore missed the extreme heat flux event described by McPhee et al. (2005), in which turbulent heat flux at 4 m exceeded 100W m^{-2} for several hours on 18 March. However, this event appears to have been quite localized, and probably not characteristic of the entire surrounding region.

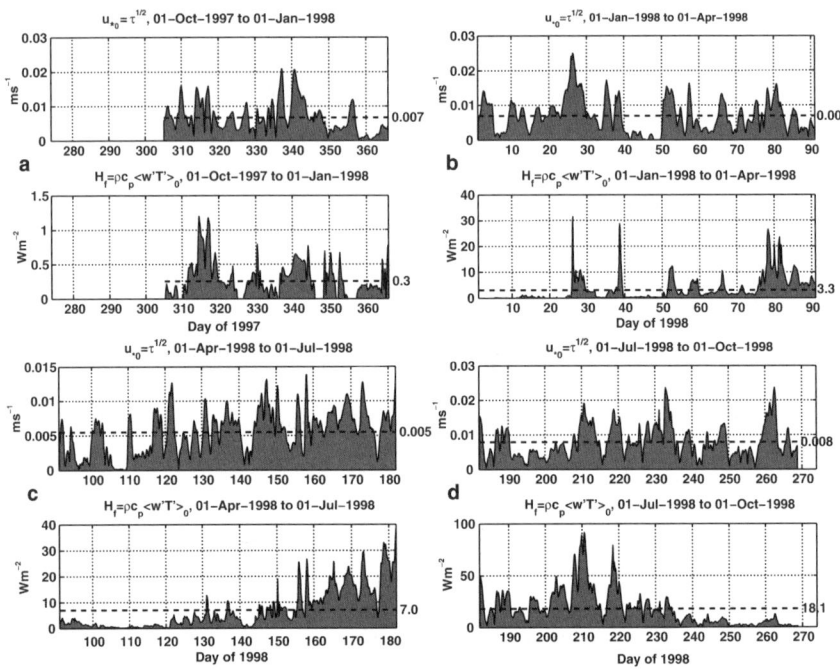

Fig. 9.13 Modeled interface friction velocity and basal heat flux by quarter during the 11-month SHEBA project. Mean values for each quarter are indicated by the labeled lines. **a** 1997Q4; **b** 1998Q1; **c** 1998Q2; **d** 1998Q3

A summary over the entire SHEBA project (Fig. 9.14) shows the mean friction velocity to be 6.7 mm s^{-1} and basal heat flux to be 7.7 W m^{-2}. The time series have been smoothed with a one-week running mean. Errors in the modeled interface flux quantities arise mainly from uncertainties in specifying z_0 and the heat exchange coefficient. In an independent study (McPhee et al. 2003), the bulk heat exchange coefficient was reported as 0.0057 ± 0.0004. For the interface model, this implies a range of $0.0088 \leq \alpha_h < 0.0102$ (when ice is melting). Similarly, based on the standard deviation of $\log(z_0)$ discussed above, a probable range for hydraulic roughness is $0.016 \leq z_0 \leq 0.146$ m. The model was run with combinations of both minimum z_0 and α_h and maximum z_0 and α_h. The corresponding ranges in u_{*0} and basal heat flux are shown by the shaded regions in Fig. 9.14. Ranges for mean values are $5.75 \leq \bar{u}_{*0} \leq 7.95$ mm s^{-1}, and $6.25 \leq \bar{H}_f \leq 9.43$ W m^{-2}. If the missing October values are estimated as the averages of September 1998, and November 1997, the most likely average annual friction velocity for the SHEBA year was 6.8 mm s^{-1} and annual average basal heat flux was 7.2 W m^{-2}. For the period 2 June 1998 to the end of the turbulence project, the most likely average heat flux was 17.6 W m^{-2}. Perovich et al. (2003) estimated the summer basal heat flux from the bottom melting observed at 77 mass balance sites scattered across many different ice types on the

9.3 Applications

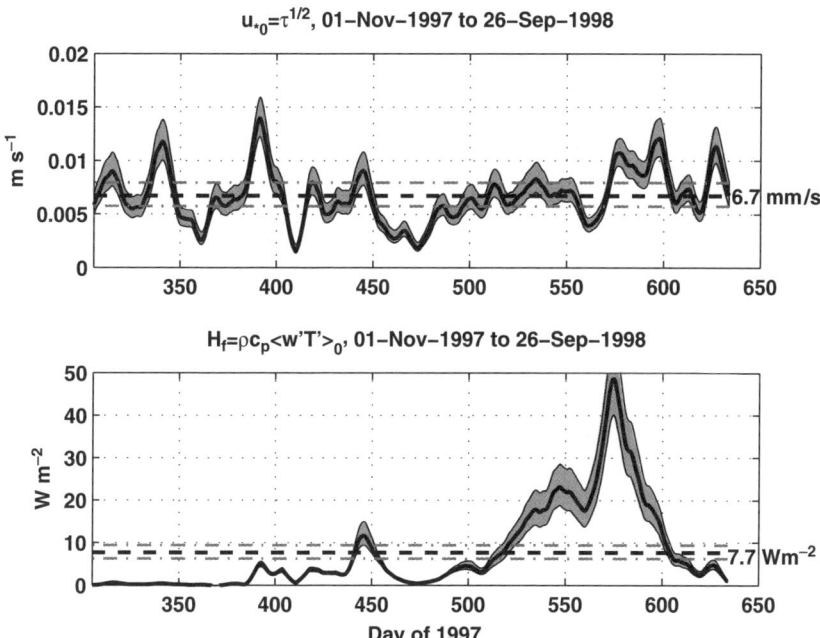

Fig. 9.14 Modeled friction velocity and basal heat flux for the entire SHEBA project, where the 3-h time series have been smoothed with a seven-day running mean. Average values for the 11 months are shown by dashed lines. Shading indicates a range of values corresponding to model calculations for the lowest combination of z_0 and heat transfer coefficient to the highest, with the range in mean values shown by the gray dot-dashed lines

SHEBA floe. For the period 2 June 1998 to 3 October 1998, they estimated the average value to be 17.5 W m^{-2}. Given the variety of assumptions underlying each estimate, the close agreement is probably fortuitous, but nevertheless adds credence to the approach. Perovich et al. (2003) report approximately equal amounts of surface and basal ablation, suggesting that about as much energy was absorbed at the surface as extracted from the ocean by melting. On the other hand, Persson et al. (2002, their Fig. 22g) estimate the total energy flux reaching the top of the ice column at SHEBA to be about 35 W m^{-2} during the summer months, June through September. Thus to balance the energy budget would apparently require that about half the energy reaching the surface of the ice would have made its way into the ocean.

At the time of writing, application of the SLTC model for deriving long time series of ice/ocean exchange is relatively new. However, preliminary comparisons between model results (from ADPs and upper ocean T/S sensors) and direct flux measurements from unmanned drifting buoys as part of the North Polar Environmental Observatory program are promising (Shaw et al. 2008, in press). With a new generation of drifting buoys equipped with profiling CTDs and ADPs, the method should provide accurate estimates of ice/ocean fluxes.

References

Andreas, E. L. and Claffey, K. J.: Air-ice drag coefficients in the western Weddell Sea 1. Values deduced from profile measurements. J. Geophys. Res., 100 (C3), 4821–4831 (1995)

Businger, J. A., JWyngaard, C., Izumi, Y., and Bradley, E. F.: Flux-profile relationships in the atmospheric surface layer. J. Atmos. Sci., 28, 181–189 (1971)

Carsey, F.: Microwave observations of the Weddell Polynya. Mon. Wea. Rev., 108, 2032–2044 (1980)

Edson, J. B., Fairall, C. W., Mestayer, P. G., and Larsen, S. E.: A study of the inertial-dissipation method for computing air-sea fluxes. J. Geophys. Res., 96, 10689–10711 (1991)

Hellmer, H. H., Schroeder, M., Haas, C., Dieckmann, G. S., and Spindler, M.: The ISPOL drift experiment, Deep-Sea Res., II, doi:10.1016/j.dsr2.2008.01.001, in press (2008)

Leavitt, E.: Surface-based air stress measurements made during AIDJEX. In: Pritchard, R. (ed.), pp. 419–429. University of Washington Press, Seattle (1980)

McPhee, M. G.: An analysis of pack ice drift in summer. In: Pritchard, R. (ed.) pp. 62–75. University of Washington Press, Seattle (1980)

McPhee, M. G.: Scales of turbulence and parameterization of mixing in the ocean boundary layer. J. Mar. Syst., 21, 55–65 (1999)

McPhee, M. G.: Turbulent stress at the ice/ocean interface and bottom surface hydraulic roughness during the SHEBA drift. J. Geophys. Res., 107 (C10), 8037 (2002), doi: 10.1029/2000JC000633

McPhee, M. G.: Physics of early summer ice/ocean exchanges in the western Weddell Sea during ISPOL. Deep-Sea Res., II, doi:10.1016/j.dsr2.2007.12.022, in press (2008)

McPhee, M. G., Kikuchi, T., Morison, J. H., and Stanton, T. P.: Ocean-to-ice heat flux at the North Pole environmental observatory. Geophys. Res. Lett., 30 (24), 2274 (2003), doi: 10.1029/2003GL018580

McPhee, M. G., Kwok, R., Robins, R., and Coon, M.: Upwelling of Arctic pycnocline associated with shear motion of sea ice. Geophys. Res. Lett., 32, L10616 (2005), doi: 10.1029/2004GL021819

Mellor, G. L. and Yamada, T.: Development of a turbulence closure model for geophysical fluid problems. Rev. Geophys., 20, 851–875 (1982)

Morison, J. H. and McPhee, M.: Ice-Ocean Interaction. In: Steele, J., Thorpe, S., and Turekian, K. (eds.) Encyclopedia of Ocean Sciences, Academic, London, doi: 10.1006/rwos.2001.0003(2001)

Perovich, D. K., Grenfell, T. C., Richter-Menge, J. A., Light, B., Tucker, W. B., III, and Eicken, H.: Thin and thinner: Sea ice mass balance measurements during SHEBA. J. Geophys. Res., 108 (C3), 8050 (2003), doi: 10.1029/2001JC001079

Perovich, D. K. et al.: Year on ice gives climate insights. Eos, Trans., Am. Geophys. Union, 80 (41), 481, 485–486 (1999)

Persson, P. O. G., Fairall, C. W., Andreas, E. L., Guest, P. S., and Perovich, D. K.: Measurements near the Atmospheric Surface Layer Flux Group tower at SHEBA: Near-surface conditions and surface energy budget. J. Geophys. Res, 107 (C10), 8045 (2002), doi: 10.1029/2000JC00705

Shaw, W. J., Stanton, T. P., McPhee, M. G., and Kikuchi, T.: Estimates of surface roughness length in heterogeneous under-ice boundary layers. J. Geophys. Res. (2008) (in press)

Skyllingstad, E. D., Paulson, C. A., Pegau, W. S., McPhee, M. G., and Stanton, T.: Effects of keels on ice bottom turbulence exchange. J. Geophys. Res., 108 (C12), 3372 (2003), doi: 10.1029/2002JC001488

Colour Plate Section

Plate 1 Photographs from a helicopter returning to the FRAM I station north of Fram Strait in March, 1979, after a radio communication informed us of a crack appearing in camp. Although a close call, no major equipment was lost. The lead continued to widen until it was about 1 km across, and the camp survived, with beachfront property for a time, and a thriving suburb to the south. The lead eventually froze hard enough to serve as the camp runway.

Plate 2 The author standing next to the upper ocean turbulence installation with the R/V *Polarstern* in the background during the 2004-2005 ISPOL project in the western Weddell Sea. At the time (late December), the ISPOL station was about 15 km due east of the track of HMS *Endurance* (Shackelton's famous drift) at the same time of year in 1915.

Plate 3 Fresh crack with a newly formed pressure ridge in the background, photographed near AIDJEX station Blue Fox, in April, 1976.

Plate 4 Aerial view of a recently formed lead in the eastern Arctic Ocean taken during a hydrographic survey based at the FRAM I station in 1979. Use of the word "lead" to describe these quasi-linear features is said to have originated with early explorers who thought they were aligned mostly north-south and hence would lead to the North Pole.

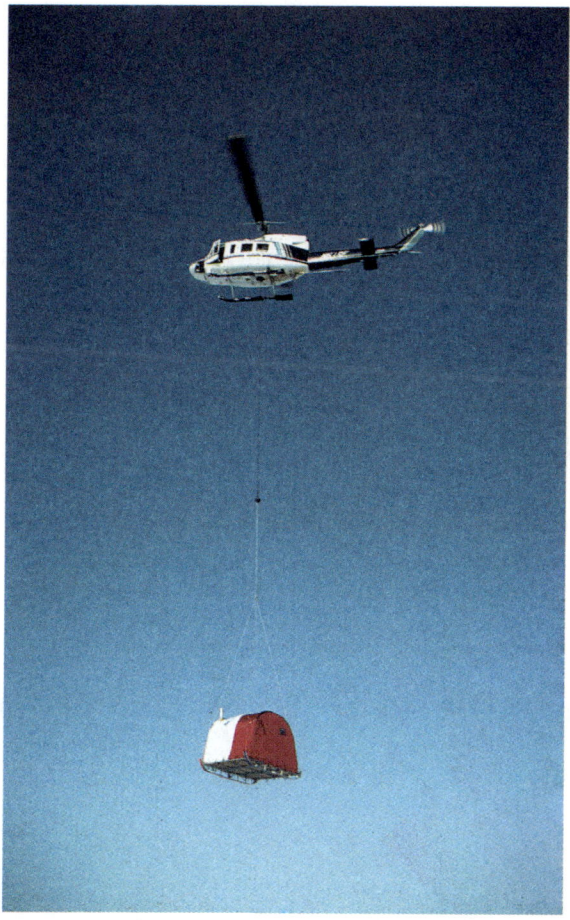

Plate 5 Just-on-time helicopter delivery of a "helo hut" to the edge of a newly opened lead during LeadEx, April, 1992. The shelter was packed with scientific gear to be deployed at the lead edge.

Plate 6 LeadEx Lead 3 temporary station, April, 1992. The station was established in a matter of hours, providing invaluable data on the impact of negative buoyancy flux from freezing on the IOBL.

Plate 7 View of the SHEBA drift station in March, 1998, from the bridge of the CCS *Des Grosielliers*, showing the lead that opened between the scientific station and the ship, temporarily disrupting most observations. Ice on the far side of the lead later shifted forward several hundred meters.

Plate 8 Allan Gill (left) preparing a CTD station at the edge of lead during the FRAM I project in 1979 north of Fram Strait, with help from the helicopter crew, Helge Siljeberg (center) and Gøran Lindmark (right). In addition to lending invaluable support during numerous scientific ice camps (and providing a role model for young researchers), Allan's Arctic experience included sledding across the North Pole from Barrow to Svalbard as a member of the British Transarctic Expedition in 1968-69.

Plate 9 Deploying a turbulence mast during the AIDJEX 1972 Pilot Study. The author, at the time a graduate student, is standing behind Prof. J. Dungan Smith. The mast was later moved into position by divers.

Colour Plate Section 201

Plate 10 Instrument tent over a hydrohole in Templefjord, Svalbard, during a UNIS student field exercise, March, 1999. Photograph courtesy of S. McPhee.

Plate 11 The National Science Foundation chartered icebreaker *R/V Nathaniel B. Palmer* moored to the ice for a short drift station during the MaudNESS project in the eastern Weddell Sea, August, 2005.

Plate 12 Emperor penguin contemplating safety floats near the turbulence mast installation during the ISPOL project in the Weddell Sea, December, 2004. The picture, taken from inside my tent shelter, shows the *R/V Polarstern* in the background.

Plate 13 Mountains and glaciers from an airplane traveling from Longyearbyen to Ny Ålesund, Svalbard, March, 2002. Photograph courtesy of S. McPhee.

Plate 14 Instrument shelter on fast ice in VanMijen Fjord, Svalbard, June, 2004.

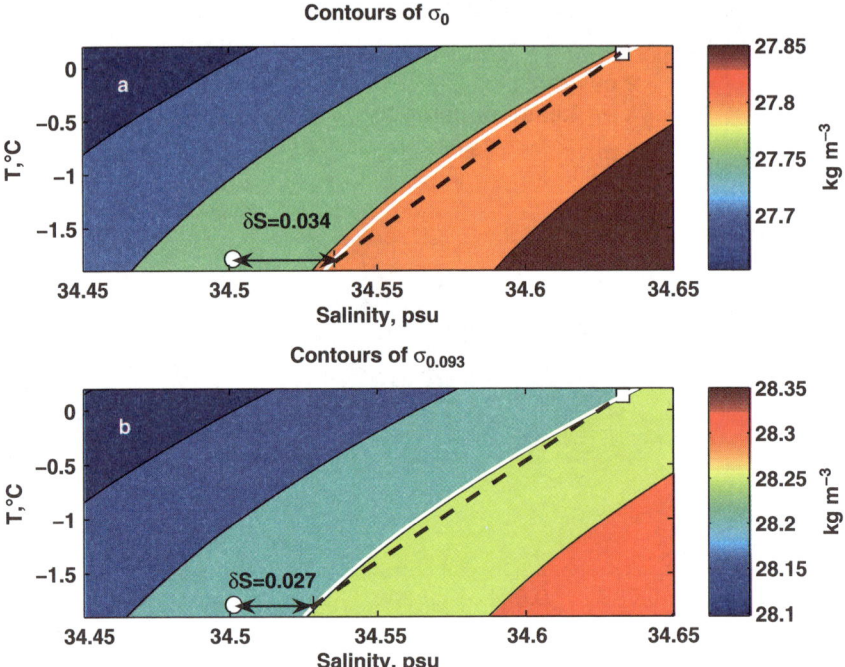

Plate 15 Temperature/salinity diagrams with isopycnal contours for density calculated at **a** surface pressure and **b** at pressure corresponding to the mixed layer depth. *T/S* characteristics of the idealized two-layer system from Figs. 2.2 to 2.10 are indicated by symbols (circle for upper, square for lower). See text for further details (see also in black-and-white on page 32)

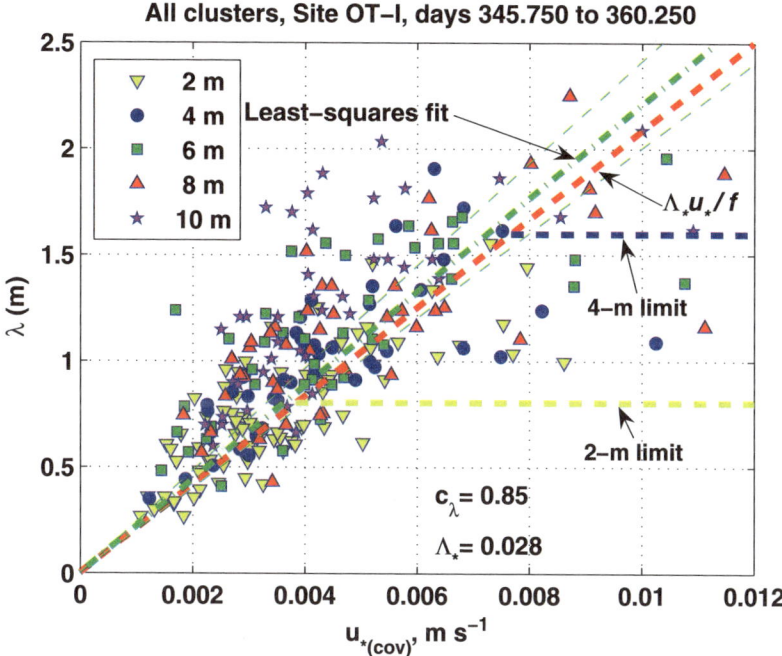

Plate 16 Scatter plot of mixing length λ versus u_* for all 3-h averages prior to the ISPOL Christmas-Day breakup. The green dot-dashed line is a least-squares linear fit through the origin with 95% confidence interval indicated by the light dashed lines. The red dashed line indicates the dynamic (planetary) maximum mixing length. The horizontal dashed lines indicate the "geometric" limits, $\kappa|z|$, at 2 and 4 m, respectively (From McPhee 2008, in press) (see also in black-and-white on page 94)

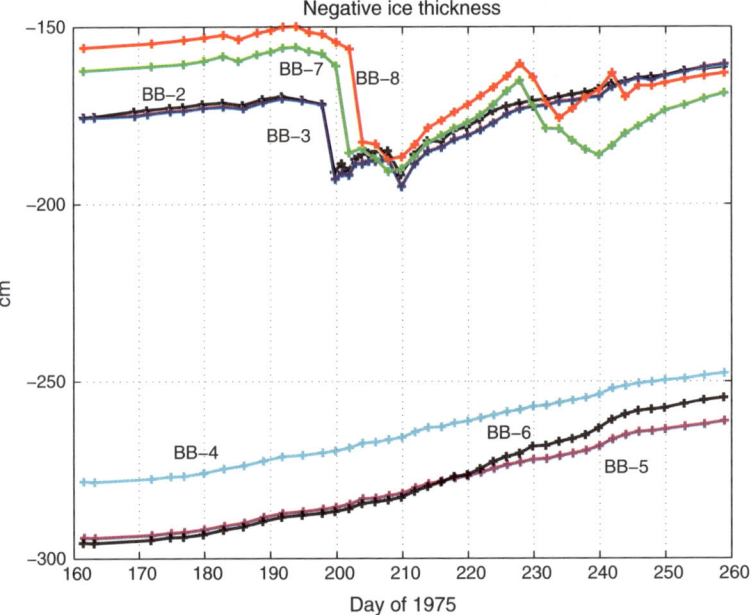

Plate 17 Ice bottom elevation relative to the upper surface from ablation measurements made by A. Hanson during the 1975 AIDJEX project in the western Arctic (Adapted from Notz et al. 2003. With permission American Geophysical Union) (see also in black-and-white on page 120)

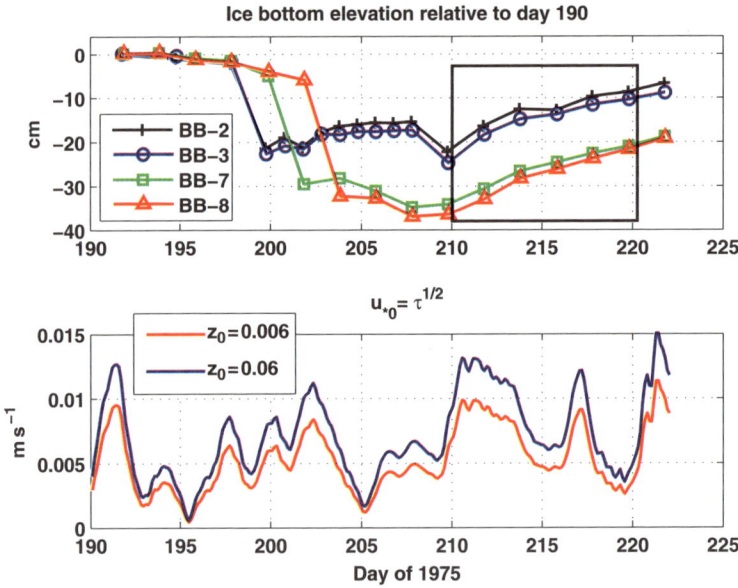

Plate 18 Upper panel: bottom elevation of "false bottom" thickness gauges relative to their readings on day 190. Box marks the ten-day period chosen for simulation. Bottom panel: Interface friction velocity determined from ice drift relative to geostrophic current, for two values of surface roughness spanning range of estimates for AIDJEX station Big Bear (Adapted from Notz et al. 2003. With permission American Geophysical Union) (see also in black-and-white on page 122)

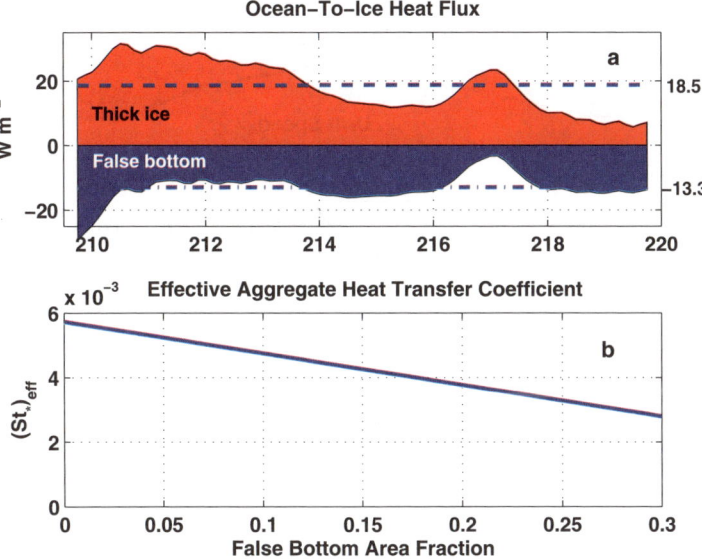

Plate 19 a Time series of heat flux to thick ice (red shading) and heat flux into the ocean from false bottoms (blue). Average values are shown at right. b Aggregate Stanton number as a function of areal coverage of false bottoms and fresh water (Adapted from Notz et al. 2003. With permission American Geophysical Union) (see also in black-and-white on page 124)

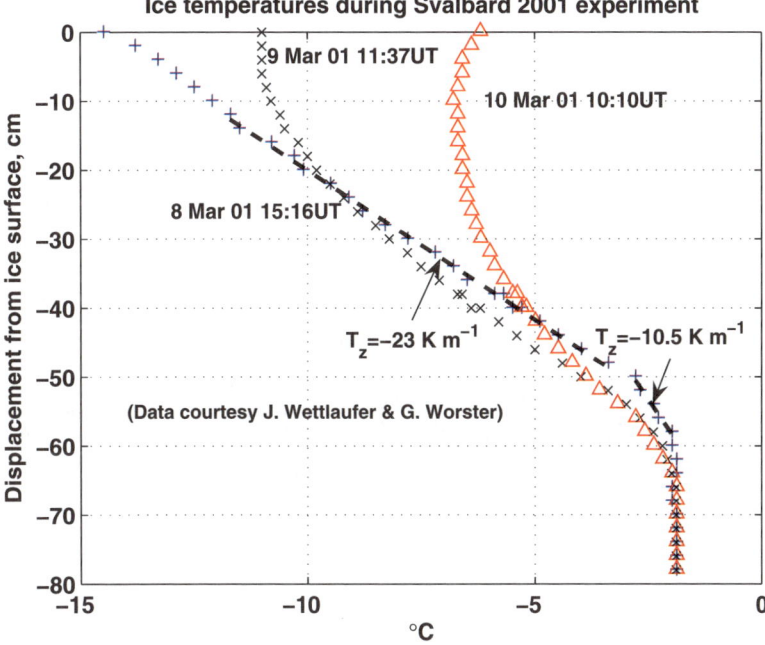

Plate 20 Ice temperature profiles on three days in fast ice on VanMijen Fjord, Svalbard (see also in black-and-white on page 128)

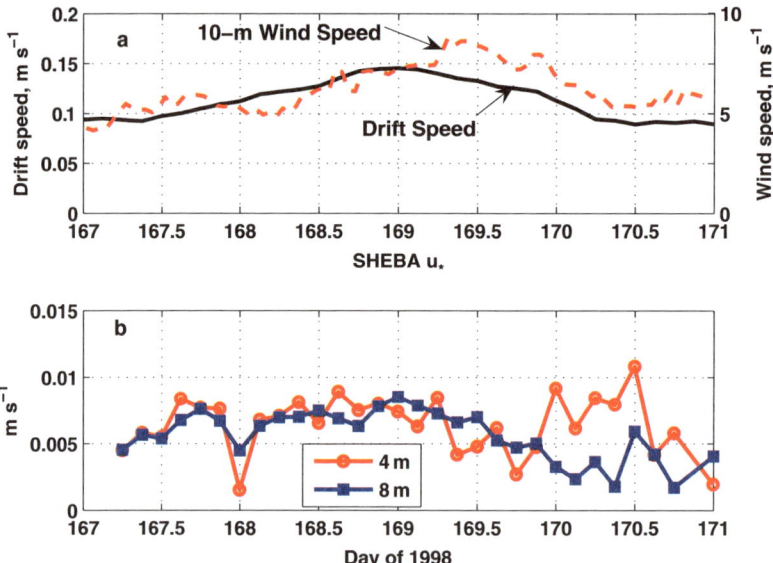

Plate 21 a Wind speed at 10 m and ice drift speed after removing inertial component, 16–20 June 1998, at the SHEBA station. Ice speed ordinate range is 2% of the wind speed range. Time is shown as days of 1998, where 167.0 is 0000UT on 16 June. **b**. Friction velocity (square root of kinematic Reynolds stress) measured at two distances from the ice/water interface (see also in black-and-white on page 146)

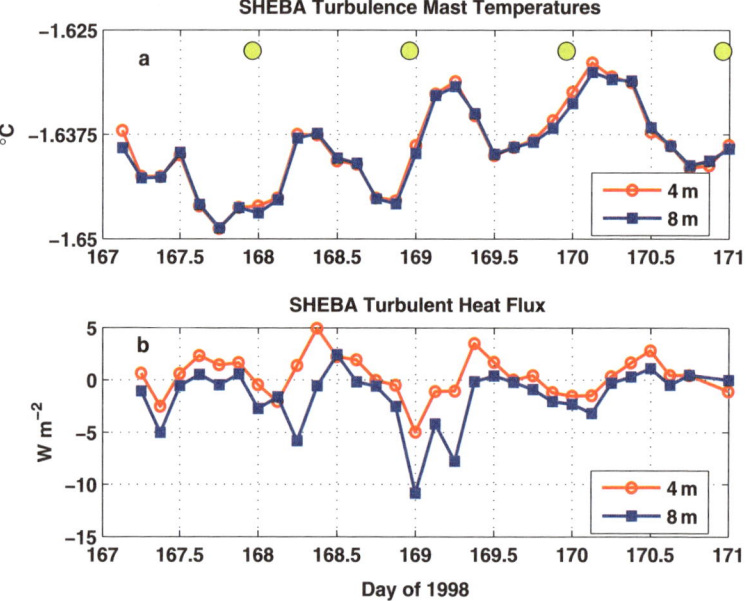

Plate 22 a Three-hour average turbulence mast temperatures. Shaded circles represent local solar zenith, at approximately UT + 23 h. **b** Corresponding heat flux measurements: $\rho c_p \langle w'T' \rangle$ (see also in black-and-white on page 147)

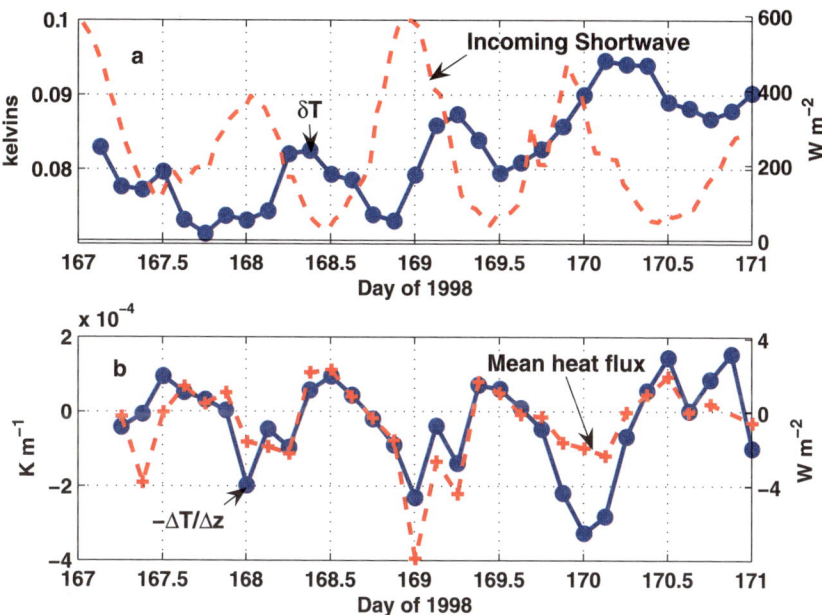

Plate 23 a Incoming shortwave radiation at the upper ice surface (data from the SHEBA Project Office installation, right caption) and departure of mast temperature from freezing (average of clusters 1 and 2 at 4.2 and 8.2 m, respectively). **b** Negative temperature gradient between clusters 1 and 2, after adjusting temperatures to agree at times near zero heat flux, along with turbulent heat flux averaged for both clusters (see also in black-and-white on page 148)

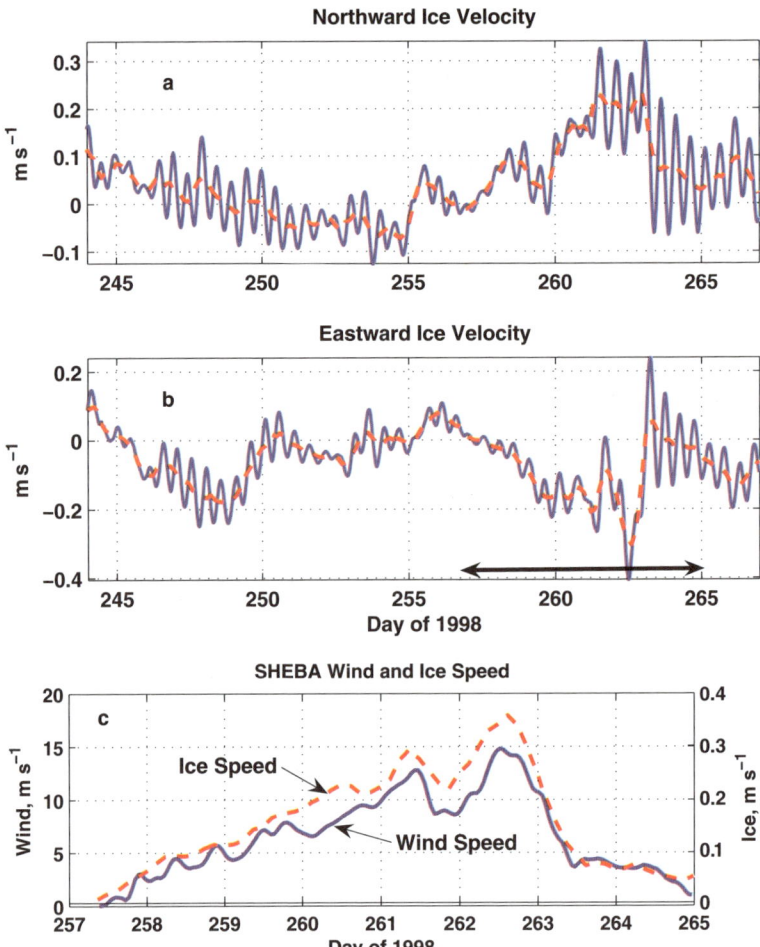

Plate 24 a Northward ice drift velocity during September of the SHEBA project from satellite global positioning data. **b** Eastward component. The arrow indicates time shown in: **c** Wind speed and drift speed during the time from 257.375 (0900 UT on 14 September 1998) to 265.0. The scale for ice speed is 2% of the wind speed scale (see also in black-and-white on page 152)

Colour Plate Section 211

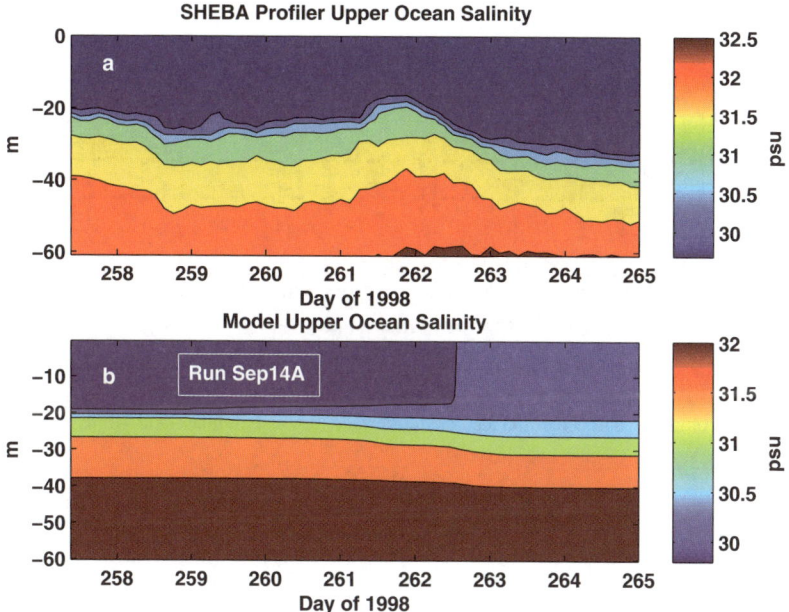

Plate 25 Contour plots of salinity in the upper ocean from SHEBA profiler **a** and wind-driven model Sep 14A **b** (see also in black-and-white on page 156)

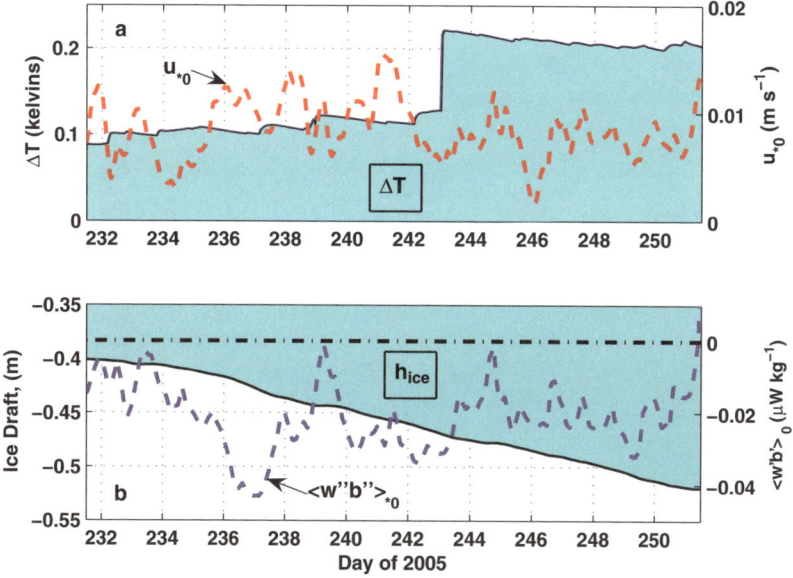

Plate 26 a PD Model interface friction velocity (dashed) and departure of mixed layer temperature from freezing (shaded); **b** ice draft (shaded) and interface buoyancy flux (dashed) for an IOBL model where all density gradients are based on potential density. (see also in black-and-white on page 165)

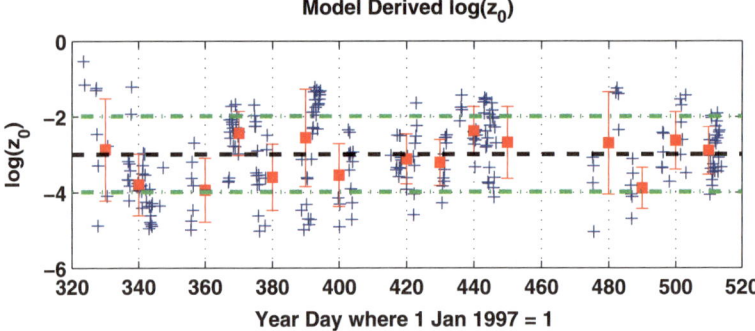

Plate 27 Time series of the natural logarithm of underice surface roughness derived from SLTC model runs based on measured Reynolds stress and relative velocity at TIC 2 on the SHEBA turbulence mast, nominally 8 m below the ice/ocean interface (see also in black-and-white on page 188)

Index

A
Acoustic-Doppler velocimeter, 43, 46
Adiabatic lapse rate, 148
Advection, 17, 19, 34, 112, 147, 155, 170
Aggregate floe roughness, 186
Aggregate Stanton number, 124, 206
AIDJEX, 2, 3, 5, 6, 10, 11, 40, 58, 70, 71, 74, 77, 87, 88, 98, 109, 110, 119, 120, 122, 124, 145, 187–189, 195, 200, 205
Angular shear, 66, 70, 186
Angular wave number, 55, 57, 63, 88
ANZFLUX, 11, 34, 117
Attenuation, 73, 74, 140, 149, 168, 180

B
Basal heat flux, 118, 119, 149, 157, 162, 165, 169, 183, 189–191
Basal melting, 25, 126, 149
Beaufort Gyre, 1, 2, 5, 40, 109, 119, 125, 187
Blasius solution, 114, 117
Bootstrap method, 46–48, 51
Boundary condition, 136, 139, 140, 148, 151, 159
Boussinesq approximation, 16
Brine volume, 111
Buoyancy
  flux, 12, 42, 44, 52, 63, 65, 68, 69, 74, 77–79, 95, 96, 98–100, 104, 105, 107, 133, 137, 141–143, 149, 165–167, 177, 179, 180, 198, 210
  frequency, 134, 141, 166, 178
  production, 52

C
Cabbeling, 33, 34
Calibration, 42, 92, 99, 102, 147
CEAREX, 11, 42, 117

Complex demodulation, 21–25, 160, 189
Confidence interval, 48–50, 56, 94, 103, 204
Congelation, 125–127
Coriolis force, 18, 19, 138, 140
Coriolis parameter, 8, 18, 37, 66
Covariance, 12, 17, 40–44, 46–51, 54, 57, 60–62, 73, 89, 91, 93, 102, 104, 127, 173, 176
Cross-spectrum, 58

D
Deep convection, 7, 162
Destabilizing buoyancy flux, 12, 44, 104, 105
Difference equation, 134, 136, 138, 139
Dimensional analysis, 53, 57, 66, 68, 72, 76, 113
Dimensionless current hodograph, 185
Dimensionless wind shear, 66
Distributed source, 140
Diurnal heating, 68, 146, 147
Double diffusion, 35, 115, 118–120, 122, 125, 127, 128

E
Eddy diffusivity, 90, 102, 133, 134, 136, 142, 144, 148, 166, 176, 178, 180, 183
Eddy viscosity, 8–10, 58, 59, 61, 67, 72–74, 76, 79, 81, 82, 87–92, 102, 133, 134, 138, 140–142, 176–180, 183, 186
e-folding depth, 140, 149
Ekman pumping, 24–28, 35
Ekman spiral, 9, 70, 73, 174
Enthalpy, 16, 109, 110
Entrainment velocity, 145
Equilibrium mass balance, 109
Extinction coefficient, 150, 151

F

False bottom, 118–125, 127, 205, 206
First law of thermodynamics, 109
Flow blockage, 174, 184
Flow chart, 141
Flux grid, 138, 158
Flux Richardson number, 78, 79, 81, 82, 86
Forced convection, 105
Form drag, 176
Fossil melt ponds, 186
Frazil, 16, 125–127, 140
Free convection, 105
Freezing point, 30, 109, 127
Friction velocity, 50, 53, 60, 66–68, 74, 77, 89, 90, 94–98, 114, 116, 117, 122, 129, 131, 136, 146, 149, 150, 152, 153, 157, 165, 166, 174–177, 181, 183, 189–191, 205, 207, 210

G

Geometric scale, 94, 95
Geostrophic current, 18, 19, 122, 135, 181, 205
Geostrophic drag coefficient, 187
Global positioning satellite, 22, 23
Grid cell, 174

H

Heat flux, 3, 7, 11, 25–27, 33, 41, 44, 45, 48, 49, 51, 54, 61, 62, 90–92, 95, 96, 102, 103, 109, 110, 113, 114, 116, 118, 119, 121, 122, 124, 127–129, 143, 145–152, 154, 155, 157–159, 162, 164, 165, 168, 169, 173, 178–181, 183, 184, 189, 190, 206–208
Heat sink, 149
Heterogeneous floe, 11, 176
Histogram, 47
Hodograph, 71, 87, 174, 175, 182–185
Hummocks, 186
Homogeneity, 59, 120
Hydraulically smooth boundary, 128
Hydraulic roughness, 180, 182, 186, 187, 190

I

Ice-albedo feedback, 3, 4, 125
Ice-edge bands, 83–85
Ice extent, 4, 7
Ice inertia, 138
Ice-ocean boundary layer, 3
Ice pump, 124, 126
Ice Station Weddell, 11, 73, 74, 87–92
Implicit solution, 133, 135
Inertial oscillation, 145, 151, 152, 157, 158, 176

Insolation, 7, 145
Interface submodel, 143, 148, 149
Interface velocity, 110, 114, 119
IOBL parameterization, 173
Isotropy, 57, 59, 88, 102
ISPOL, 11, 70, 93, 94, 174–176, 184, 185, 187, 194, 202, 204
Iterative scheme, 176, 177

K

Kolmogorov constant, 57, 61, 63, 89

L

Laser-Doppler velocimeter, 43
Latent heat, 109–112, 121
Law of the wall, 67, 128, 129, 137, 181
LeadEX, 11, 42, 44, 50, 58, 98–100, 102–105, 107, 142, 197, 198
Leapfrog, 135, 140
Light transmittance, 151
Local friction velocity, 89, 90, 177
Local turbulence closure, 12, 65, 128, 133, 140, 145, 176

M

Marginal ice zone, 11, 41, 67, 85, 95, 96, 113, 114, 116–119
Marginal static stability, 162
MaudNESS, 12, 43, 46, 49, 146, 162–164, 170
Mean quantity grid, 137, 181
Melt ponds, 124, 125, 149, 186
Microstructure conductivity sensor, 42, 45
Microstructure profiler, 102
Mixed-layer evolution, 145
Mixing length, 56, 58, 67, 70, 74, 78, 87, 89–91, 94, 101, 102, 133, 141, 142, 177, 184, 185, 187, 204
MIZEX, 11, 41, 84, 85, 95–99, 113, 114, 116, 117, 126
Monin-Obukhov similarity, 59, 65, 68, 70, 107
Mushy layer, 129

N

Navigation, 21–23, 161, 180, 182, 183
Nocturnal cooling, 147, 149
Nonlinearities, 31, 33, 146, 167, 168
Normal probability distribution, 48
Numerical model, 12, 68, 128, 133, 145, 174

O

Obukhov length, 69, 79–81, 83, 99, 100, 137, 141
Outer layer, 70, 72, 80, 82, 90, 94, 95, 104

Index

P
Partial differential equation, 133
Percolation velocity, 110, 115, 143
Planetary scale, 66, 72, 76, 80, 83, 94, 95, 141
Potential density, 26, 31–33, 162, 165–167, 210
Potential temperature, 91, 92, 102, 148, 177, 178, 183
Practical salinity scale, 16, 30, 111
Prandtl number, 113–115
Pressure ridge keel, 27, 40, 59, 174, 176, 183, 187
Pycnocline, 3, 5, 26–28, 35, 77, 80, 106, 107, 141, 142, 145, 154, 155, 158, 160, 162, 165, 167, 178, 180, 189

Q
Quadratic drag, 68, 77, 102, 112, 187

R
Radar, 160
Recursion relation, 135, 139, 140
Reynold's analogy, 142
Reynolds flux, 16–18, 173
Reynolds number, 39, 40, 53, 61, 66, 113, 114, 117
Reynolds stress, 17, 18, 52, 53, 59, 66, 73, 87, 91, 94, 104, 135, 146, 152, 178, 180, 181, 187, 188, 207, 211
Richardson number, 78, 142, 145, 178
Rossby similarity, 28, 65, 74–78, 81, 83, 121, 185, 188
R/V Nathaniel B. Palmer, 11, 12, 43, 201
R/V Polarstern, 11, 194, 202

S
Saline contraction factor, 30, 31, 37, 143
Salinity flux, 42, 44, 99, 100, 112, 113, 118, 127, 129
Salt balance, 110–112
Schmidt number, 113–115
Sea-Bird Electronics, 41, 99
Seasonal pycnocline, 2, 77, 107
Shear production, 53, 58, 59, 78, 95, 105
SHEBA, 1–3, 11, 21, 22, 24–29, 35, 36, 42, 43, 52, 59–61, 68, 110, 117, 119, 122, 125, 145–152, 155, 156, 160, 162, 163, 174, 177, 178, 183, 186–191, 198, 207–211
Slab model, 145
Smith rotors, 40–41
Solar
 nadir, 147
 radiation, 3, 16, 102, 109, 116, 124, 125, 145, 149, 151, 184
 zenith, 44, 146, 147, 207

Specific heat, 16, 110
Spectral gap, 17, 48–51
Staggered grid, 133, 134, 139
Stanton number, 114, 116–118, 122–124, 127, 206
Steady local turbulence closure, 173–191
Steady-state solution, 139, 176
Stratification, 7, 28, 83, 85, 94, 158, 160, 178
Streamline coordinate system, 88
Stress spiral, 73
Supercooling, 125–127
Surface friction Rossby number, 76
Surface layer, 34, 52, 59, 65–70, 72–76, 78, 80–82, 89, 94, 101, 104, 107, 112, 142, 173, 177, 181, 182

T
Taylor's hypothesis, 17, 50
Thermal barrier, 33, 162, 166, 168
Thermal conductivity, 111, 164
Thermal dissipation, 54, 60, 89, 102
Thermal expansion factor, 30, 143
Thermal gradient, 90, 92, 102, 104, 111
Thermal mixing length, 89, 102
Thermal variance production, 61, 102
Thermal wind, 19, 135
Thermobaricity, 33, 34
Thickness gauge, 119, 122, 205
TKE dissipation, 56, 89, 95
Topographic effects, 185
Turbulence instrument cluster, 41–43
Turbulent eddy, 16, 17, 48, 50
Turbulent exchange coefficient, 112–114
Turbulent kinetic energy, 12, 18, 51–53, 89
Turbulent realization, 49, 61
Turbulent scale velocity, 104, 117, 141–143
Turning angle, 68, 75, 77, 78, 187

U
Underice morphology, 70, 152, 173
Upwelling, 9, 25–28, 34, 35, 142, 154

V
Van Mijen Fjord, 43, 127
Variance spectrum, 87
Volume transport, 20, 24, 70, 145
Von Karman's constant, 66

W
Warm deep water, 7, 146
Wave radiation, 85
Weddell Polynya, 7, 33, 34